QING COLONIAL ENTERPRISE

LAURA

Qing
En

Ethnogra
Ear

The University of Chicago Press, Chicago 60637
The University of Chicago Press, Ltd., London
© 2001 by The University of Chicago
All rights reserved. Published 2001
Printed in the United States of America
25 24 23 22 21 20 19 18   4 5 6

ISBN-10: 0-226-35420-2 (cloth)
ISBN-13: 978-0-226-35421-7 (paper)

Library of Congress Cataloging-in-Publication Data

Hostetler, Laura.
    Qing colonial enterprise : ethnography and cartography in early modern China /
Laura Hostetler.
        p.   cm.
    Includes bibliographical references and index.
    ISBN 0-226-35420-2 (alk. paper)
        1. Ethnology—China.   2. Cartography—China.   3. China—Colonization.   4. China—
Social life and customs.   5. China—History—Ch'ing dynasty, 1644–1912.   I. Title.
    GN635.C5 H67 2001
    951'.03—dc21

                                                                              00-010974

For Beulah and John,
for blazing trails without burning bridges

and for Naomi

CONTENTS

CHAPTER SEVEN

**The Evolution of a Genre: Miao Albums as Art
and Objects of Study**

CONCLUSION

NOTE ON ROMANIZATION

The *pinyin* system of romanization is used throughout the text. Titles of publications and proper names normally written in other forms of romanization have not, however, been altered to conform to the *pinyin* system. Other exceptions are made in regard to Peking (rather than Beijing), Taipei, and the Yangzi River. For the sake of consistency other forms of romanization within quoted passages have been altered to *pinyin*, but appear within brackets to indicate the change.

TABLES, MAPS, FIGURES, AND COLOR PLATES

Tables

Maps

FIGURES

COLOR PLATES
Following page 108

PREFACE AND
ACKNOWLEDGMENTS

his book aims to reach several audiences: historians and students of the Qing empire (1636–1911);[1] historians of the early modern period (broadly construed as 1500–1800) who, while specializing in other geographic locations, are concerned with processes of empire building worldwide; and those with expertise in the history of ethnography and cartography in other parts of the world interested in a comparative case.

In building its empire the Qing state made extensive use of technologies of representation often associated exclusively with the early modern period in Europe. Although its situation was in many ways unlike that of Europe (the Russian empire might provide more obvious parallels), several compelling reasons lead me to this comparison. First, the history of early modern Europe is a common point of reference for most of the intended audience of the book. Secondly, and more importantly, a comparison with Europe challenges assumptions made about the fundamental differences between "East" and "West." This study provides historians of early modern Europe with a case study of practices in another part of the world that have heretofore been seen as largely unique to the West—and as such responsible for its ascendancy during the latter part of the early modern period. At the same time, it takes issue with the still too prevalent myths of Chinese isolationism and Chinese exceptionalism by demonstrating explicitly how Qing practices of mapping both territory and peoples were in many ways comparable with those used by

---

1. I follow Pamela Crossley in dating the Qing from its founding rather than from 1644 when it extended its rule to succeed the Ming.

European colonial powers. My choice of early modern Europe and Qing China as points of comparison stem from my background and research abilities, not a desire to privilege the importance of these two regions over others. I hope that more scholars will take up the challenge of comparative work on the question of early modernity in other world areas.

Thanks are due to many people and institutions who helped to make this work possible. First and foremost to my dissertation advisor, Susan Naquin, who urged me to discover and pursue my own interests and who, by challenging me to convince her of the validity of my findings, gave me the confidence that I could convince others too. Thanks also for support and encouragement from Victor H. Mair, whose broad professional interests all relate in one way or another to breaking down the myths of Chinese isolationism. David Buisseret, James Cracraft, Beulah S. Hostetler, David P. Jordan, Mark Liechty, Marion S. Miller, Evelyn S. Rawski, William T. Rowe, and Joanna Waley-Cohen read parts or all of the manuscript, or dissertation out of which it grew, and made suggestions for improvements. I am extremely grateful also to Edwin J. Van Kley and to Richard J. Smith, both of whom shared their considerable expertise with me in reviewing the manuscript for the press. Any remaining errors and inconsistencies are solely my responsibility.

For institutional support I would like to thank the following. The Institute for the Humanities at the University of Illinois at Chicago provided full-time support for writing during 1997–98, and made possible a visit to the British Library in 1999. The National Endowment for the Humanities provided a summer research stipend (1997) and a dissertation fellowship (1993). The German Academic Exchange Service (DAAD) financed a research trip to Germany to study Miao albums. Thanks also to the Newberry Library, where much of the research on cartography was carried out with the helpfulness of Jim Akerman, Director of the Hermon Dunlap Center for the Study of Cartography, and Robert Karrow and Pat Morris, Curator and Assistant Curator of the Newberry's map collection. Portions of the manuscript appeared in *Modern Asian Studies* 34.3 (2000): 623–62, under the title "Qing Connections to the Early Modern World: Ethnography and Cartography in Eighteenth-Century China." I am grateful to Cambridge University Press for permission to reprint this material here in somewhat altered form.

I owe a debt of gratitude to the staffs of many libraries and museums in the United States and abroad in which Miao albums are now housed. These include, in the United States: the National Museum of Natural History, Smithsonian Institution, Washington, D.C.; the Gest Oriental Library, Princeton University; the Harvard-Yenching Library, Harvard University; the Library of Congress; the American Museum of Natural History, New York; the New York Public Library; and the University Museum, Philadelphia. In Britain: the Bodleian Library, Oxford University; the British Library, London; and the Pitt-Rivers Museum, Oxford University. In France, the Musée Guimet (Musée National des Arts Asiatiques). In Germany: the Forschungs- und Landesbibliothek, Gotha; the Museum für Völkerkunde, Berlin; and the Museum für Völkerkunde, Leipzig. In Italy: the Italian Geographical Society, Rome; and the Vatican Library. In the People's Republic of China: the National Library, Beijing; Beijing University Library; the Central Minorities Institute, Beijing; the Chinese Academy of Social Sciences, Peking; the Guizhou Provincial Museum; and the Guizhou Provincial Library. In Taiwan, R.O.C.: the Fu-ssu Nien Library, Institute of History and Philology, Academia Sinica. I am grateful also for the use of the Regenstein Library at the University of Chicago and the guidance of their excellent bibliographic staff.

I would like to express my appreciation to members of the faculty of Guizhou Normal University who arranged the details of my stay at that institution and scheduled visits with other scholars from the area working on local history and minority studies. Special thanks to Schuyler Jones of the Pitt-Rivers Museum for his introduction to this group of scholars, to Heather Peters for introductions in Peking, and to Sören Edgren for calling my attention to the albums in the American Museum of Natural History, New York.

Ray Brod provided invaluable assistance with providing the sketch maps, and Wang Yi with preparation of the glossary. Agostino Cerasuolo and Benjamin Tam bore the brunt of grading and logistics for my survey course on Chinese history from 1500–1911 while the manuscript was in the final stages of preparation. Kim Pilcher provided the quiet hours needed for sustained concentration by caring for Naomi.

I would also like to thank my parents, John A. and Beulah Stauffer Hostetler, and my in-laws, Russel A. and Marjorie S. Liechty, for their unfailing support in both tangible and intangible ways. Over the years their households have been a haven for their children, extended families, and friends, who have found nourishment of many kinds around their tables. These overlapping communities, which include readers, critics, cooks, counselors, musicians, writers, editors, gardeners, listeners, poets, scholars, skeptics, visionaries, teachers, parents, and children, have made me who I am today. I only hope that in the future I will be able to extend the table by creating circles similar in spirit wherever the vagaries of time may take me. Finally, thanks to Mark Liechty, my partner in the process. Without his support this book would not have been written.

INTRODUCTION

Cartography and Ethnography as Early Modern Modes of Representation

his book explores the Qing state's use of cartographic and ethnographic representation in the building of empire. How are Qing representations of territory and the peoples who populated it similar to those made in various countries of Europe during the early modern period? Why is this an important question? Simultaneous developments in cartographic and ethnographic modes of representation notable for their emphasis on empirical knowledge derived from direct observation and precise measurement suggest that the Qing was not isolated from global changes during the early modern period, nor was it simply a recipient of European knowledge; it was an active participant in a shared world order.

In defining "early modern" I draw on the work of Sanjay Subrahmanyam, who would decouple the often assumed link between "European" and "early modern." He argues that this period "represents a more-or-less global shift, with many different sources and roots." Among its central defining features are "momentous changes in conceptions of space and thus cartography" as well as the emergence of "significant new empirical 'ethnographies.'"[1] New ways of conceptualizing the world mirrored in new technologies of representation were a product of exploration not only by European countries, but by centers of power worldwide. Qing overland imperial expansion at once contributed to, and

1. Sanjay Subrahmanyam, "Connected Histories: Notes towards a Reconfiguration of Early Modern Eurasia," *Modern Asian Studies* 31.3 (1997): 737. Subrahmanyam's definition allows us to get away from the idea of early modern as a purely political and economic step on a teleological path to "modernity." See Jack Goldstone, "The Problem of the Early Modern World," *Journal of the Economic and Social History of the Orient* 41 (August 1998): 249–84.

benefited from, these same early modern cartographic and ethnographic modes of representation. The state employed both to extend its claims to Universal Empire—another feature of the early modern period to which Subrahmanyam points.[2] In short, during the late seventeenth and eighteenth centuries Qing China was more fully a part of what can be called the early modern world than has been generally recognized.[3]

This book contributes to the debate on the early modern non-West, arguing that the term "early modern" can appropriately be used to describe global, rather than uniquely Western, processes. We need to see the Qing during this period as part of, rather than isolated from, the early modern world. In its use of various modes of visual representation, the Qing began to view, and constructed images of, its peoples and territory in ways that can best be described as early modern. Using both

2. For aspirations to Universal Empire under the Qianlong emperor, see Pamela Crossley, *A Translucent Mirror: History and Identity in Qing Imperial Ideology* (Berkeley and Los Angeles: University of California Press, 1999). For work on the third feature of early modernity mentioned by Subrahmanyam, that of conflict between "settled agricultural and urban societies . . . and nomadic groups" ("Connected Histories," 738), see Peter Perdue, "Boundaries, Maps, and Movement: Chinese, Russian, and Mongolian Empires in Early Modern Central Eurasia," *International History Review* 20.2 (June 1988): 263–86.

3. Chinese historians of this period are beginning to take a more outward-looking approach. See Nicola Di Cosmo, "Qing Colonial Administration in Inner Asia," *International History Review* 20.2 (June 1988): 287–309; Valerie Hansen, *The Open Empire: A History of China to 1600* (New York and London: W. W. Norton and Company, 2000); Laura Hostetler, "Qing Connections to the Early Modern World: Ethnography and Cartography in Eighteenth-Century China," *Modern Asian Studies* 34.3 (2000): 623–62; Peter C. Perdue, "Military Mobilization in Seventeenth and Eighteenth-Century China, Russia, and Mongolia," *Modern Asian Studies* 30.4 (1996): 757–93, and "Boundaries, Maps, and Movement"; Kenneth Pomeranz, *The Great Divergence: Europe, China, and the Making of the Modern World Economy* (Princeton: Princeton University Press, 2000); Joanna Waley-Cohen, "China and Western Technology in the Late Eighteenth Century," *American Historical Review* 98.5 (1993): 1525–44, "Commemorating War in Eighteenth-Century China," *Modern Asian Studies* 30.4 (1996): 869–99, "Religion, War, and Empire-Building in Eighteenth-Century China," *International History Review* 20.2 (June 1988): 336–52, and *The Sextants of Beijing: Global Currents in Chinese History* (New York and London: W. W. Norton and Company, 1999); and Roy Bin Wong, *China Transformed: Historical Change and the Limits of European Experience* (Ithaca, N.Y.: Cornell University Press, 1997). See also Michael Adas, "Imperialism and Colonialism in Comparative Perspective," *International History Review* 20.2 (June 1988): 371–88, and the other articles in the same issue of the *International History Review*, which is devoted to Manchu Colonialism. For a fresh approach to Qing foreign relations, see James Hevia, *Cherishing Men from Afar: Qing Guest Ritual and the Macartney Embassy of 1793* (Durham and London: Duke University Press, 1995). For works dealing with Asia more broadly, see Donald F. Lach and Edwin J. Van Kley, *Asia in the Making of Europe*, vol. 3, *A Century of Advance*, bk. 4, *East Asia* (Chicago: University of Chicago Press, 1993); James D. Tracy, ed., *The Rise of Merchant Empires* (Cambridge and New York: Cambridge University Press, 1990), and *The Political Economy of Merchant Empires* (Cambridge and New York: Cambridge University Press, 1991). For a review of additional literature on the early modern world see the Introduction to Andre Gunder Frank's *Reorient: Global Economy in the Asian Age* (Berkeley: University of California Press, 1998).

Chinese and European primary documents, I examine specific instances of the development and use of ethnography and cartography to demonstrate that the techniques of expansion that the Qing employed, and the epistemology behind these techniques, were similar to those that shaped early modern European expansion. This should not be surprising, for the circumstances that required their implementation for successful state building stemmed from the same geopolitical imperatives. These included, but were not limited to, a recognition of the world's finite area as the globe became charted according to both latitude and longitude, and massive population growth that encouraged state expansion and accounted for increased contact across cultures.

CARTOGRAPHY

The modern term for "map," whether in Romance, Slavic, Chinese, Indian, or even premodern Middle Eastern languages, derives from broader terms relating either to the material on which drawings were made (*mappa*, Latin for cloth; *chartes*, Greek for papyrus), or to pictures or illustrations more generally (premodern Middle Eastern languages,[4] Indian languages,[5] Russian,[6] and Chinese). The specific term "cartography," referring to the drawing of what we now think of as a map — as distinct from other representations of territory and their inhabitants — was not coined until 1839.[7] The modern Chinese word for map is *ditu*, literally a picture, or illustration, of the earth or of land. In classical Chinese the term *tu* alone is often employed. However, *tu* can mean not only map but as a noun can also refer to an illustration, a plan, a chart, or a painting. As a verb it can mean to plan, to scheme, or even to covet. Thus our modern definition of map as a scaled image of territory has narrowed considerably over time. As discussed below, maps

4. Nathan Sivin and Gari Ledyard, "Introduction to East Asian Cartography," in *The History of Cartography*, vol. 2, bk. 2, *Cartography in the Traditional East and Southeast Asian Societies*, ed. J. B. Harley and David Woodward (Chicago and London: University of Chicago Press, 1994), 27.

5. P. D. A. Harvey, *The History of Topographic Maps: Symbols, Pictures, and Surveys* (London: Thames and Hudson, 1980), 10.

6. According to James Cracraft, "The Old-Russian word for 'map'—*chertëzh*—meant any rough drawing or sketch and was used in various contexts, notably in building." During the first half of the eighteenth century, "the general European word for map was naturalized in Russian—as *karta*." James Cracraft, *The Petrine Revolution in Russian Imagery* (Chicago: University of Chicago Press, 1997), 273–75.

7. Josef Konvitz, *Cartography in France, 1660–1848: Science, Engineering, and Statecraft* (Chicago: University of Chicago Press, 1987), xix.

including ethnographic images of peoples were common in sixteenth- and seventeenth-century Europe.

While mapping in China may be nearly as old as Chinese civilization itself, Qing pursuit of accurate, to-scale, cartographic representation of the empire and its frontiers is contemporaneous with similar developments in early modern Europe. We know that the development of cartography in Europe paralleled the growth of national consciousness and an era of exploration abroad.[8] Similar processes taking place in seventeenth- and eighteenth-century China are only beginning to be examined.[9] Contextualizing the mapping of the Qing within the larger (international) early modern project of geographical learning allows us to see that the Qing dynasty was more closely linked to the early modern world than heretofore recognized.

Between 1708 and 1718 the Kangxi emperor (r. 1662–1722) commissioned a team of European Jesuit missionaries in his service to survey and map the extent of his empire. The resulting maps, which appeared in Chinese, Chinese and Manchu, and various European language versions, were different from other contemporary Qing maps in that they were drawn to scale, and as such required no accompanying text detailing distances from one location to another. The Kangxi emperor's desire for "a precise map which would unite all the parts of his empire in one glance"[10] corresponded roughly with Peter the Great's mapping of Russia, French cartographic projects at home and in the New World, and early British colonial exploits in India. This convergence in mapping activity, techniques, and even in the network of specific historical figures involved, can best be explained as independent yet interrelated responses to global conditions similarly affecting these world powers.

8. Benedict Anderson, *Imagined Communities: Reflections on the Origin and Spread of Nationalism*, revised edition (London and New York: Verso, 1991), especially chapter 10, "Census, Map, Museum"; David Buisseret, ed., *Monarchs, Ministers, and Maps: The Emergence of Cartography as a Tool of Government in Early Modern Europe* (Chicago and London: University of Chicago Press, 1992); Alfred W. Crosby, *The Measure of Reality: Quantification and Western Society, 1250–1600* (Cambridge: Cambridge University Press, 1997), especially chapter 5, "Space."

9. Joanna Waley-Cohen and Theodore Foss have convincingly argued that the Kangxi emperor was well aware of the uses and value of cartographic accuracy, that he was not simply a passive player unwittingly allowing outsiders access to information about his realm. Waley-Cohen, "China and Western Technology"; Theodore N. Foss, "A Jesuit Encyclopedia for China: A Guide to Jean-Baptiste Du Halde's *Description . . . de La Chine . . .* (1735)," (Ph.D. diss., University of Chicago, 1979).

10. Joseph-Anne-Marie de Moyriac de Mailla, *Histoire générale de la Chine; ou, Annales de cet empire: Traduites du Tong-kien-kang-mou*, vol. II (Paris, 1780; reprint, Taipei: Ch'eng-Wen Publishing Company, 1967), 314. Translations are by the author unless otherwise indicated. Cf., Peter Perdue, "Boundaries, Maps, and Movement," 263–86.

The qualities that have come to characterize early modern mapping in Europe are not intrinsically European per se; they are as much a product of the historical period during which they emerged as the location. This distinction might seem simplistic, or even obvious, but it is too often overlooked. Examining specific historic examples in the development of cartography from both Europe and China will help to recast our categories of thought from the too easily assumed dichotomy between "Chinese" and "European," (or an even more facile "East" and "West") to think instead along the lines of premodern vs. early modern forms of representation identifiable in both world regions. The historic differentiation of what I will call geography—which often described both peoples and territories in the same documents—into the distinct areas of inquiry now called ethnography and cartography can be traced in both China and Europe.

ETHNOGRAPHY

Mapping was not the only form of visual representation that underwent qualitative shifts during the early modern period in both the Qing empire and parts of Europe. Representations of the inhabitants of territories being mapped also exhibit a notable transformation. To demonstrate this transition I examine the collection and categorization of ethnographic knowledge as reflected in Qing dynasty texts about its southwestern frontier populations, focusing particularly on the residents of Guizhou province. These texts include local gazetteers and histories, but also colorful illustrated manuscripts known in English as Miao albums.[11] A careful look at the development of ethnographic texts in local gazetteers and histories from the early seventeenth to the mid-eighteenth century shows that the quest for knowledge about non-Chinese peoples on the empire's internal frontiers, carried out by official representatives of the Qing state, was increasingly characterized by the rigor of direct observation and empirical method.

The album genre arose during the first half of the eighteenth century

---

11. The generic name in Chinese is *Miao man tu,* or *Bai Miao tu.* The term derives from the general expression *miao man,* rendered loosely as "southern barbarian," and does not refer specifically to China's Miao ethnic groups related to the Hmong of present-day Southeast Asia.

The central focus on Guizhou province grows naturally out of my primary sources. Of the more than eighty albums I examined, over seventy-five percent were devoted to groups living within Guizhou. Illustrated albums of minority groups did exist and do survive in lesser numbers for other provinces, including Yunnan, Taiwan, Hunan, and Sichuan, and when relevant I also refer to them.

in the context of Qing colonization of frontier areas where military expansion and population growth led to increased contact and conflict between Han Chinese settlers, the majority ethnic group, and their non-Han neighbors.[12] The texts of the albums are largely based on information found in the gazetteers. The illustrations depict courtship practices, religious customs, seasonal festivals, and sketches from everyday life. The albums were initially designed to educate officialdom about the habits and customs of various non-Han ethnic groups in order to be able to govern them more effectively. In their manner of collecting, organizing, and conveying information, they suggest a degree of familiarity with political technologies that harness the acquisition of "scientific" knowledge to the goals of the state. Indeed, a shift in ways of viewing the world and thinking about what constitutes knowledge was central to the ethnographic enterprise in late Ming (1368–1643) and Qing China as it was in early modern Europe. I use the term "ethnography" to distinguish these emerging records of "other" peoples that were based on direct observation from earlier descriptions based on legend and hearsay. Increasingly accurate depictions of peoples and territory had become crucial aspects of early modern empire building—both for the Qing and in Europe.

THE EARLY MODERN SHIFT IN MODES OF REPRESENTATION

What kinds of questions and assumptions about cartography have we inherited? How have these notions framed the questions with which we approach both Chinese mapping practices and their interaction with foreign technologies in the eighteenth century? Where are the weak spots in the current analysis of eighteenth-century joint Sino-European cartographic projects? What might be a more productive way of

12. Han is a term with a long history. In the eighteenth century the term *hanren* (Han people) referred to those people who considered themselves descendants of the Han dynasty (202 B.C.– 220 A.D.). The term *hanren* was also used to distinguish culturally Chinese peoples from their non-Han neighbors (i.e., Miao, Zhuang, etc.). Since the late nineteenth or early twentieth century the term *han* has come to refer to anyone who is, appears, claims, or is assumed to be, ethnically Chinese. Sun Yat-sen and other republican revolutionaries found it a useful construct for unifying the population (north and south) against the Manchus. For an in-depth study of the modern term *han* as a construct, see Leo J. Moser, *The Chinese Mosaic: The Peoples and Provinces of China* (Boulder and London: Westview Press, 1985). At a broad level of generalization *"han"* is a bit like the term "white." It brings together a diverse grouping of people under one common rubric, thus creating a shared condition that can exist only by virtue of its contrast to apparently greater difference—"black," Hispanic, Manchu, Miao, etc.

approaching and understanding their significance? These questions are closely interrelated and need to be addressed at the outset of this work.

One of the most recent, and in some ways most comprehensive, accounts of mapping in China is found in the multivolume *History of Cartography* edited by David Woodward and Brian Harley.[13] With the exception of the "Introduction to East Asian Cartography" by Nathan Sivin and Gari Ledyard and one chapter on "Chinese Cosmographical Thought" by John Henderson,[14] Cordell D. K. Yee authored all of the chapters on China. The overarching theme that connects his essays on a wide variety of topics and periods is the argument that too narrow a conception of what constitutes a map, and a preoccupation with tracing technological advances in cartography, have led to a lack of appreciation for and understanding of indigenous Chinese mapping practices. As he makes clear, premodern Chinese maps were not monolithic in their design and execution, but of a wide variety. The 1136 *Yu ji tu* (Map of the Tracks of Yu) provides an example of an early Chinese map that was made to scale (figure 1).[15]

Prior to Yee's series of articles in the *History of Cartography,* the most

13 Harley and Woodward, *Cartography in the Traditional East and Southeast Asian Societies* (see note 4 above). See also Cao Wanru, et al., eds., *Zhongguo gudai ditu ji,* vol. 3, *Qing dai* (Beijing, 1997); Richard J. Smith, *Chinese Maps* (Hong Kong, Oxford, and New York: Oxford University Press, 1996), and "Mapping China's World: Cultural Cartography in Late Imperial Times," in *Landscape, Culture, and Power in Chinese Society,* ed. Yeh Wen-hsin (Berkeley: University of California, Center for East Asian Studies, 1998)"; Lothar Zögner, *China cartographica: Chinesische Kartenschätze und europäische Forschungsdokumente* (Berlin: Staatsbibliothek Preussischer Kulturbesitz, 1983); Benjamin Elman, "Geographical Research in the Ming-Ch'ing Period," *Monumenta Serica* 35 (1981–1983): 1–18; Hostetler, "Qing Connections to the Early Modern World"; Li Xiaocong, *A Descriptive Catalogue of Pre-1900 Chinese Maps Seen in Europe* (Peking: Guoji wenhua chubangongsi, 1996); James A. Millward, "'Coming Onto the Map': 'Western Regions' Geography and Cartographic Nomenclature in the Making of Chinese Empire in Xinjiang," *Late Imperial China* 20.2 (December 1999): 61–98; Mark Elliott, "Mapping the Qing State: *All* Under Heaven?" (paper presented at the annual meeting of the Association for Asian Studies, April 6–9, 1995, Washington, D.C.); and four papers presented at the annual meeting of the Association for Asian Studies, March 26–29, 1998: C. Patterson Giersch, "Mapping an Imperial Frontier into 'National Territory': How Qing Officials Demarcated the Yunnan-Burma Frontier and Helped Produce a Corner of China"; John Herman, "Mapped and Re-mapped: Chinese Representations of the Shuixi Region during the Seventeenth Century"; Laura Hostetler, "Representation and Empire: Mapping the Qing as a Colonial Enterprise"; and Emma Teng, "Mapping Emptiness: Visual and Literary Representations of Taiwan's Wilderness."

14. John B. Henderson, "Chinese Cosmographical Thought: The High Intellectual Tradition," in Harley and Woodward, *Cartography in the Traditional East and Southeast Asian Societies,* 203–27.

15. See Harley and Woodward, *Cartography in the Traditional East and Southeast Asian Societies,* 46–48; and Joseph Needham, "Geography and Cartography," in *Science and Civilisation in China,* vol. 3 (Cambridge: Cambridge University press, 1959), 547–48.

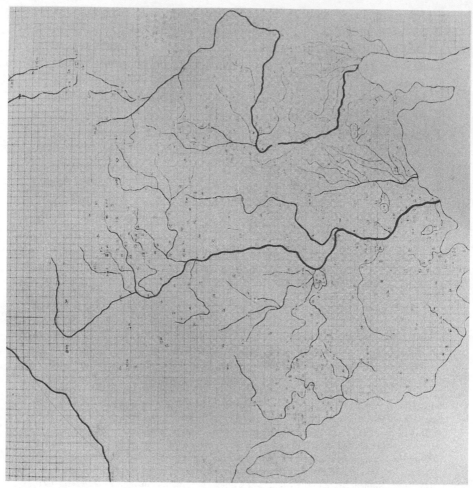

Figure 1 *Yu ji tu*. Map of the Tracks of Yu, 1136. Reprinted with the permission of Cambridge University Press.

comprehensive work on cartography in China was Joseph Needham's extensive chapter on the subject, to which Yee's work implicitly responds.[16] Looking at the arguments of these two authors provides a sense of the parameters of the debate on cartography in China and how it has been shaped by a comparison to cartographic practices in the West. Needham's thorough familiarity with scientific traditions in both China and Europe gives his work a strong foundation. Although differ-

16. Needham, "Geography and Cartography," 497–590.

ing in approach and interpretation, his work bears some similarity to Yee's if only in that both have tried to dispel any sense that "Chinese" cartography was somehow inferior to "Western" cartography. Whereas Yee's method is to open up a broader conception of what constitutes a map to enlarge the scope of our appreciation, the heart of Needham's method was rather to show that cartographic practice in China was not inferior to that practiced in Europe even as judged by modern standards of accuracy.[17]

The premise of most of Yee's work is that we need to get away from arguments based on assumptions about the superiority of one kind of cartography over another. While I concur with his argument that our concept of "map" has been too narrowly constructed to allow for a full appreciation of mapping practices in China, the portion of Yee's work dealing with Qing interaction with early modern Europe is somewhat dissatisfying, perhaps in part because it is confined largely to the same framework or question that his predecessors addressed; i.e., does mapping in China become increasingly "scientific" over time? Even while resisting this teleological view, he is drawn into a problematic, although well-established, dichotomy that has dominated thinking on mapping practices in China. That is the opposition between "Chinese" and "European" (or "Western"), which corresponds (too) neatly with "traditional" vs. "modern" or "scientific." Because he falls into this conceptualization he sees the Kangxi atlas as primarily a "European" work employing technology that never really caught on in China.[18] He finds additional evidence in the fact that maps combining text and nonscaled images continued to be prevalent in China well into the nineteenth century.

We need, I think, to synthesize the most valid points of both Needham and Yee while at the same time reaching beyond the parameters of their arguments to pose and attempt to answer a different set of questions. Our options open up when we re-frame the debate. As long as we assume a sharp dichotomy between what is "Chinese" and what is "European," or "Western"—a holdover from the earlier literature that Yee perpetuates in his "Traditional Chinese Cartography and the Myth of

17. Needham's chapter is not, however, confined to a narrow conception of what constituted cartography in China. He includes, for example, sections on "Anthropological Geographies" and "Descriptions of Southern Regions and Foreign Countries," in which he gives a history of ethnographic depictions of "other" peoples, including the early *zhigong tu* (Illustrations of Tributaries). See pp. 508–14.

18. Cordell D. K. Yee, "Traditional Chinese Cartography and the Myth of Westernization," in Harley and Woodward, *Cartography in the Traditional East and Southeast Asian Societies*, 170–202.

Westernization"[19]—comparison and contrast cannot be avoided. "Traditional" Chinese maps, we learn, tend not to be drawn to scale, include a great deal of text, and are sometimes pictorial. While this broadens our understanding of what a map can be, it also tends to lock Chinese mapping into a normative model; i.e., if it is not pictorial and does not include textual description it is not a (traditional) Chinese map. An equally serious weakness in this logic revolves around unspoken assumptions about what "Western" cartography is, and that Western cartography constitutes a viable and discreet entity coterminous with "scientific" cartography. Too often we tend to conflate "Western" with what is rather modern, or early modern. While Yee's careful research illuminates the broad range of representation that can, and should, be categorized as mapping in China, his point of comparison with "Western" cartography is not thoroughly historicized and thus remains artificially narrow. During the mid-Qing a number of kinds of mapping practices, reflecting various epistemologies, did coexist. Distinct technologies and map styles were suited to different audiences and purposes.[20]

The uniqueness of traits described as traditionally Chinese (not to scale, pictorial, and relying heavily on text) breaks down when one examines the history of European cartography. At one time mapping in Europe, like "traditional" Chinese mapping, was more pictorial, not drawn to scale, and did not portray firm national boundaries. Furthermore, it often had a cosmological basis in which scale and "accurate" depiction of the natural world, and of political boundaries, was not particularly valued. The schematic T-in-O map of the world, which occurred in many variations and continued to be produced from the middle ages into the seventeenth century, is simply the most obvious case in point (figure 2).[21] These important points are sometimes overlooked because works on comparative cartography tend to examine its historical development in one world area in detail, while treating the comparative case as static or unchanging. In work dealing with cartography during the Qing and earlier periods of Chinese history, "European" cartography as a point of comparison is often reduced to its

19. Yee, "Traditional Chinese Cartography."

20. John Henderson notes that during the seventeenth century agreement surfaced that the models that had been in place for cosmographical mapping since the Han dynasty did not fit with current realities. A uniform view on how to solve the problem did not emerge; a number of epistemological visions competed. Henderson, "Chinese Cosmographical Thought.".

21. For more on medieval maps in Europe, see David Woodward, "Reality, Symbolism, Time, and Space in Medieval World Maps," *Annals of the Association of American Geographers* 75.4 (1985): 510–21.

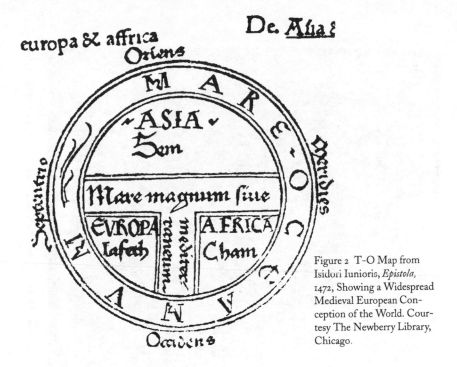

Figure 2 T-O Map from Isidori Iunioris, *Epistola*, 1472, Showing a Widespread Medieval European Conception of the World. Courtesy The Newberry Library, Chicago.

scientific qualities and/or its related usefulness as a tool in colonial conquest.[22]

A close association between what we now refer to distinctly as cartography and ethnography was common in maps made in Europe during the sixteenth and seventeenth centuries. These maps often included illustrations of people, and even pictorial scenes of the countryside, as shown in reproductions from two sixteenth-century atlases (plate 1 and figure 3). A different variation of the same admixture of cartographic and ethnographic information can be found on a 1590 map of Virginia printed by Theodore de Bry (figure 4). Looking closely, one sees, near the coastline, depictions of a warrior and of an indigenous woman with her child. Larger prints of these figures, accompanied by descriptive text, were bound into the same volume that contains the map (figure 5).[23]

22. I have heard James Akerman make this point on several occasions. See, for example, Yee, "Traditional Chinese Cartography." This is also true of Thongchai Winichakul's groundbreaking work *Siam Mapped: A History of the Geo-Body of a Nation* (Honolulu: University of Hawaii Press, 1994).

23. Thomas Hariot, *A Briefe and True Report of the New Found Land of Virginia*, printed at Frankfort in 1590 by Theodore de Bry (facsimile edition, Manchester: The Holbein Society, 1888).

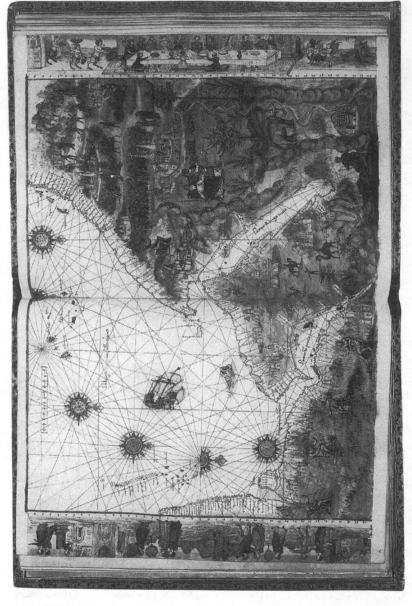

Figure 3 Map of East Africa, Arabia, and the West Coast of India, from the Atlas of Nicolas Vallard, 1547. The Henry E. Huntington Library, San Marino, Calif.

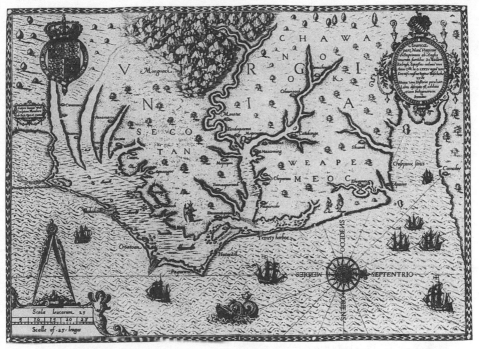

Figure 4 Map of Virginia, Reproduced from an 1888 Facsimile of Thomas Hariot's *A Briefe and True Report of the New Found Land of Virginia,* Printed at Frankfort in 1590 by Theodore de Bry. Facsimile edition, Manchester: The Holbein Society. Courtesy The Newberry Library, Chicago.

In sixteenth-century maritime atlases made in European countries, it is common to see indigenous peoples pictured in their own environment, often on the edge of the known world. Coastlines were marked "cartographically" but the interior of unfamiliar lands was depicted more pictorially. Different peoples, customs, and habitats were worth learning about because they were unfamiliar; these lands and their peoples were a curiosity. In a sense it was precisely their difference that made their pictorial representation of interest and value on the map.

In seventeenth-century European world maps we continue to see an interest in the customs and costumes of other peoples, but their classification is more schematic or rigid. Dutch world maps commonly included costumed figures along the borders, as seen in a map printed by Claes Janszoon Visscher in 1614 (figure 6). Cornelis Danckerts's 1651 edition of Petrus Plancius's world map (figure 7) includes not only costumed figures on the borders, but extensive text (often considered a

A cheiff Ladye of Pomeiooc. VIII.

bout 20. milles from that Iland, neere the lake of Paquippe, ther is another towne called Pomeioock hard by the fea. The apparell of the cheefe ladies of dat towne differeth but litle from the attyre of thofe which lyue in Roanaac. For they weare their haire truffed opp in a knott, as the maiden doe which we fpake of before, and haue their fkinnes pownced in thefame manner, yet they wear a chaine of great pearles, or beades of copper, or fmoothe bones 5. or 6. fold obout their necks, bearinge one arme in the fame, in the other hand they carye a gourde full of fome kinde of pleafant liquor. They tye deers fkinne doubled about them crochinge hygher about their breafts, which hange downe before almoft to their knees, and are almoft altogither naked behinde. Commonlye their yonge daugters of 7, or 8. yeares olde do waigt vpon them wearinge abowt them a girdle of fkinne, which hangeth downe behinde, and is drawen vnder neath betwene their twifte, and bownde aboue their nauel with mofe of trees betwene that and thier fkinnes to couer their priuiliers withall. After they be once paft 10. yeares of age, they wear deer fkinnes as the older forte do. They are greatlye Diligted with puppetts, and babes which wear brought oute of England.

Figure 5 "A cheiff Ladye of Pomeiooc," Reproduced from an 1888 Facsimile of Thomas Hariot's *A Briefe and True Report of the New Found Land of Virginia,* Printed at Frankfort in 1590 by Theodore de Bry. Facsimile edition, Manchester: The Holbein Society, 1888. Courtesy The Newberry Library, Chicago.

Figure 6 World Map, Claes Janszoon Visscher, Amsterdam, 1614. Top and bottom show pairs of people from different countries. The sides show views of twenty cities. Badische Landes-bibliothek.

hallmark of "traditional" Chinese maps) that describes, among other things, the inhabitants of various regions of the world. So too did the world map in Chinese made by Matteo Ricci in 1602 (figure 8). Maps printed in various Blaeu atlases also included right- and left-hand borders showing pairs of figures representative of various countries depicted on the maps. The top borders show views of famous cities (figure 9).

By the eighteenth century these more pictorial elements, as well as descriptive text, had disappeared from European maps. What does it mean when the inhabitants of territory disappear from maps? Why do we no longer expect to find people depicted along with territory? I would suggest that this shift in modes of representation in cartographic practice is related to new ways in which the state came to be conceptualized during the early modern period. Both in the formation of modern

Figure 7 World Map, Petrus Plancius—Cornelis Danckerts, Amsterdam, 1651. Department of Maps, Prints, and Photographs, the Royal Library, Copenhagen.

nation-states and in expansion abroad, territory came to be viewed more and more as a resource to be dominated or controlled by a political center. The land took on a value separate and distinct from those who occupied it. Furthermore, the eventual creation of nations was often predicated on an ideology of sameness. There was no reason to remind map viewers of differences among the population of "their" nation, or even that territory colonized as part of growing empires was home to other peoples. New maps, whether of places as close to home as Normandy on French domestic maps, or as far away as India on British colonial maps—reflected this attitude. Ethnography, as discussed at length below, developed as its own area of inquiry, but its place was no longer on the map.

The transition from maps including proto-ethnographic elements to scaled maps devoid of text and pictures was not as thoroughgoing in

Qing China. Pictorial maps relying on text, not scale, to communicate distances constitute the overwhelming majority of Qing maps.[24] Nonetheless, similar trends can be identified on certain types of maps. The *Huangyu quanlan tu* (Map of a Complete View of Imperial Territory) commissioned under the Kangxi emperor and later revisions commissioned under the Qianlong emperor share qualities consistent with eighteenth-century European maps. Like their European counterparts they are devoid of pictorial elements. The maps are based on astronomical points used to calculate latitude and longitude, and accordingly are drawn to a precise scale. Furthermore, unlike the majority of earlier Chinese maps, there is no accompanying text beyond the labeling of place names. With to-scale cartography based on astronomical surveys, texts detailing the exact distance between various points became, like pictorial illustrations, superfluous. For the sake of simplicity I refer to the *Huangyu quanlan tu* as the Kangxi atlas.[25]

China had no monopoly on "traditional" maps, nor did "scientific" cartography develop uniquely in early modern Europe. Benjamin Elman and John Henderson have both commented on changing epistemologies in China during the seventeenth century that affected the way Chinese literati thought about geography. Elman draws a connection between *kaozheng*, or "evidential learning," a late Ming and early Qing branch of scholarship that valued the weighing of empirical evidence in order to make sound scholarly judgments, and the practice of geography in China:

> As research pushed forward in the seventeenth century, geography became a key discipline. Despite the internal turn of this research away from concern with foreign lands, achievements in geographical knowledge during this period were most evident in the areas of military defense and historical and descriptive geography. Such achievements lent themselves to the cumulation of geographical knowledge. Cumulative progress was possible because evidential scholars, building on the efforts of their predecessors, stressed an empirical epistemology and focused on research topics that allowed for a sense of the continuity of geographical research. As a result, geography emerged as a precise discipline during the Ming-[Qing] transition period.[26]

24. Cordell D. K. Yee, "A Cartography of Introspection: Chinese Maps as Other Than European." *Asian Art* (fall 1992): 25–47.

25. The *Huangyu quanlan tu* is commonly referred to in other sources as the "Kangxi Jesuit atlas." I prefer "Kangxi atlas" because not only Jesuits were involved in the project.

26. Elman, "Geographical Research in the Ming-Ch'ing Period," 17–18. See also John B. Henderson, *The Development and Decline of Chinese Cosmology* (New York: Columbia University Press, 1984), and Henderson, "Chinese Cosmographical Thought."

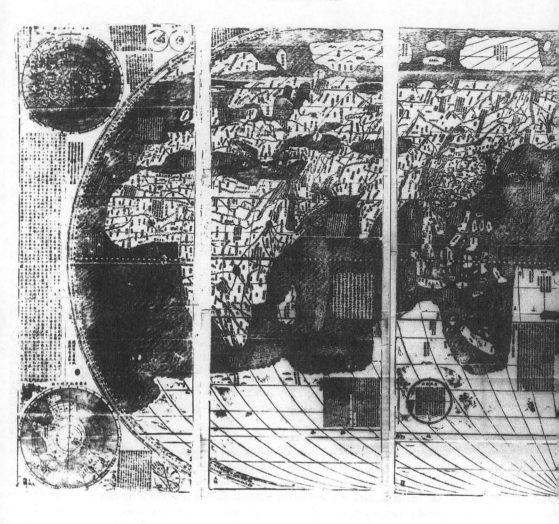

By looking at indigenous historical developments, Elman reaches beyond the dichotomy that is easily created when comparing developments across cultures to show that "[t]he empirical emphasis of evidential studies had implications for Confucian statecraft that began to be felt a century before the forced introduction of western technology into China."[27]

The practice of cartography, whether in China or Europe, does not follow a given trajectory toward some kind of predetermined most

27. Elman, "Geographical Research in the Ming-Ch'ing Period," 16.

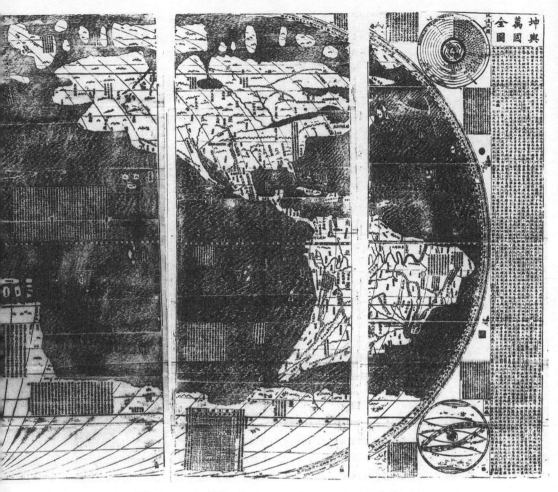

Figure 8 Matteo Ricci's 1602 Map of the World. Sotheby's.

desirable outcome. Yet, like Elman, who described how *kaozheng* scholars in the Qing built on the research of their predecessors, I am unwilling to divorce process and historical development entirely from the
practice of cartography. While a single straightforward trajectory of
"progress" in cartography is neither historically accurate nor necessarily
a desirable model for most periods, we need to recognize that certain
regions will exhibit like practices over time (at least within certain map
genres) and that the criteria that the mapmakers and map users employ
will evolve in a way that is based on a historical process of continuity,

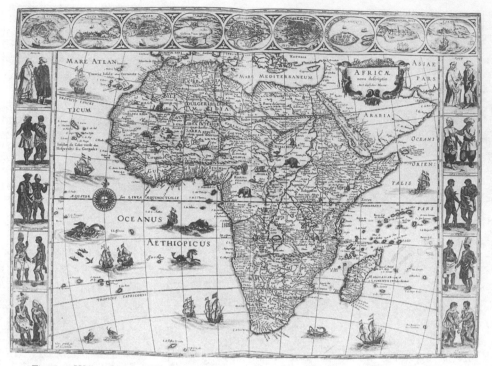

Figure 9 Willem Janszoon Blaeu's Map of Africa, from *Le théâtre du monde*, Amsterdam, 1646–1650. Courtesy The Newberry Library, Chicago.

naturally building on past experience and new knowledge (of whatever epistemological orientation). Thus, even while denying a model of linear progress, I find it valid to speak of an early modern period that can be characterized by its method, the time period in which it can be identified, and by its reach across various societies or civilizations.

Invoking this broader historical context in the study of maps allows us a fresh perspective. In Yee's view, "the standardization of map scale in the Jesuit atlas . . . [which] allowed the maps in the atlas to stand independent of text . . . was another departure from Chinese tradition."[28] I would claim that it was not so much a departure from *Chinese* tradition, as a departure from premodern to early modern forms of representation. A similar transition occurred in Europe, but we do not express the transformation in terms of where the new technology came from. It would be possible to trace the international routes through which Ptolemy's maps were preserved and transmitted, and to delve into the

28. Yee, "Traditional Chinese Cartography," 185.

channels whereby new ideas entered Renaissance Europe from outside, contributing to the renewal and transformation of European civilization. However, instead of searching for geographical origins of the concepts that later contributed to the rise of early modern mapping in Europe, we describe the shift as one that reflects the advent of a new epoch. To insist on referring to maps produced in China with technologies adopted from outside as "Western," rather than being open to the possibility that they could be both Chinese (or Manchu) and early modern, is to fall prey to assumptions about the unchanging nature of Chinese "culture" as opposed to a dynamic history claimed by "the West." The continued use of more "traditional" maps in local gazetteers and many other map genres in preference to scaled maps well into the nineteenth century does not mean that the Qing court could not or did not value early modern techniques of representation where they were useful. A multiplicity of types of cartographic representations can and inevitably does coexist in any culture at any given point in time.

Much more meaningful than a taxonomy that pits what is "Chinese" against what is "European," or "Western," is one that can embrace these differences within a framework large enough to account for differences and similarities within them over time. I would like to suggest that while national, or cultural, cartographies will continue to have distinctive qualities, it is much more useful to think about cartography as an increasingly international enterprise where the more meaningful division is between indigenous, early modern, and modern—as judged by the kinds of technologies and priorities employed in making, and subsequently reflected by, the map.

According to this definition indigenous maps are almost entirely autochthonal in their outlook and composition. The system of meaning they encode and in which they speak would not necessarily be interpretable by persons outside of the language and cultural community. They might be based on either a cosmological or a more practical geographical basis, but recognition and use would be limited to the society by which they were created.[29]

29. I choose "indigenous" rather than "premodern" here because such maps are not necessarily related to stages of development, and are not in any way confined to "primitive" peoples, or to a certain time period. They are rather designed to function within a relatively closed system of meaning. What I am calling "indigenous maps" exist aplenty in American society today. Most of us are familiar with the poster map that depicts a New Yorker's view of the world. Those who "get it" or find it amusing share a common perception of New Yorkers' perceptions of the world. Someone from China, despite training in reading modern maps, might well find it incomprehensible, not being in on the joke.

Early modern maps, by contrast, are characterized by the rhetoric of science and exact measurement. Their value and interpretation depend precisely on a reliance on accurate physical measurement and representation based on what we call scientific principles. Maps developed during the early modern period were unique because they relied on a system of communication which had come to *transcend* national languages and closed cultural systems. This kind of science may have developed into its recognizable form in European countries, but it can in no way be confined to any given nation, region, or culture. The very reliance on the graticule for determining the shape of the world, which enabled the shift in mapping to what I call "early modern," also made the pursuit of geographical knowledge an international endeavor. The quality of science that has allowed it such success in shaping the modern world is its very transcendence of political and cultural boundaries and its demands for the allegiance of any entity that would like to be considered as part of what we call the modern world. Science and technology transcended earlier cultural boundaries and in turn began to transform dominant cultural values in many parts of the world.

At the same time that the use of the graticule internationalized the study of geography, it also led to making national boundaries both more important and more distinct. Once all of the earth's surface could be plotted and accounted for, the finite nature of the land on the globe became apparent in a much more graphic way. States needed to stake out their territory against competitors. Hence the global move toward the current political system predicated on the modern nation-state. With the development of distinct national boundaries, earlier systems of interstate relations involving tribute and the possibility for multiple sovereignties or allegiances gradually became outmoded conceptually.

By the eighteenth century, thinking about territory in terms of how it fit onto a finite globe representing the earth's surface had become commonplace for cartographers using astronomical points to calculate the distance between fixed points on the earth. Consequently the earth's landmass was perceived as a fixed quantity. As empires and emergent nations vied for their portion of the pie (or melon) they saw no reason to let slices go unclaimed. Similarly no piece could be shared by two powers. Multiple sovereignties possible earlier, where small states may have paid tribute to more than one neighboring country, became conceptually impossible as maps gradually came to demarcate fixed borders *between* states.[30] Those world powers that wanted to participate in the race for territory and to stake claims that would be recognized by others would need to play the mapping game by the new rules.

I am somewhat less concerned here with defining the "modern map" since my main period of inquiry is the eighteenth century. But a few words are in order. Generally speaking there are three points that need to be made. First, the "modern map" of which I will speak does not exist to the exclusion of other kinds of maps, but it is rather the dominant kind of map found in modern societies. Second, the "modern map," depending on the scale, tends to be predicated on the system of the modern nation-state. For example, on international maps different nations are most often shown in different colors, the demarcation of national boundaries being one of the primary purposes of such maps. (Hence also the cartoons found in such publications as *Newsweek* soon after the breakup of the Soviet Union in which cartographers were kept frantically busy by the changing political shape of the former Soviet Union). The primary purpose of the modern map is to depict the political boundaries of the sovereign nation-state. This priority in large part reflects changes initiated during the early modern period, in which national boundaries (while not unchanging) became fixed.

Finally, the third quality of a modern map is that it can be interpreted by anyone trained in reading the genre. Specific languages are only important in as far as marking place names. However, the logic of the map (i.e., scale), what kinds of things are represented, and by what general types of symbols, have become almost universally interpretable. Because of these shared conventions of the modern map an international visitor can find his or her way around a strange city on public transportation without knowing the local language.

The gradual specialization of geographic representation into the distinct disciplines of ethnography and cartography is related to modern rather than specifically European ways of thinking about space and peoples. This book examines both types of representation, arguing (1) that in mapping the territory of the expanding empire, the Qing court purposely chose to use the same idiom, or map language, in which its competitors functioned; and (2) that, as with much early modern European ethnography, Qing ethnography was also directed toward use in governance of an expanding empire.

30. Thongchai, *Siam Mapped*, 55–56, 88, 94, 101.

The question of whether these developments occurred independently or whether the Qing adopted techniques introduced by the West is complex. As demonstrated in later chapters, parallels in the emergence of ethnographic depiction based on direct observation and the increasing importance placed on mapping territory can initially be traced independently in both regions. With the arrival of European Jesuits during the late Ming, geographical knowledge and other kinds of information began to be exchanged. As a result, Europe's picture of Asia gradually grew more complete, and Chinese literati were introduced to the concept of the scaled world map. As ties became closer through increased Jesuit presence at the Qing court, it was only natural that the leaders of this expanding empire would look outward to keep abreast of and patronize cutting-edge developments in the technologies employed by other expanding kingdoms.

In the eighteenth century the Qing court patronized scientists and technicians from Europe. New techniques were adopted for specific purposes. They did not supplant older, more familiar practices, but were confined to the specific contexts in which they proved most useful. At this time France, Russia, and Qing China all looked to employ the most capable scientists and cartographers to more accurately map the earth's surface; they all employed international personnel in this effort. The national origin of these employees was not a major issue; their skills and loyalty to the court by which they were employed were. Only later as these mapping techniques came to pervade European countries and the United States have we begun to think of them as "Western." We need to distinguish between making use of techniques introduced by Westerners and adopting "Western" technology.[31] During the eighteenth century techniques of to-scale mapping and the use of latitude and longitude to locate points on the earth in relation to each other were not "Western" per se, but technologically advanced. Only a few scientists in a handful of countries were familiar with these techniques. The Qing court, like other early modern states, chose to adopt these cutting-edge technologies for state building. The Qing kept up with, and arguably exceeded, the accomplishments of France and Russia until the well into the eighteenth century. More useful than asking if China adopted

31. See Martin W. Lewis and Kären E. Wigen, *The Myth of Continents: A Critique of Meta-Geography* (Berkeley: University of California Press, 1997), 49–53, for a critique of the term "Western" and an explanation of what it has meant at different times and in different contexts. While I recognize the term as problematic, I use it here nonetheless in the sense of Europe and North America, but do so in part to refute some of the logic the term has come to entail.

"Western" technology is the question of how Qing China engaged cutting-edge technology of the early modern period to position itself as a major world power.

Historiography and Implications

A number of reasons account for past resistance in both Chinese and Western historiography to considering Qing expansion in a comparative context with early modern Europe. For many Chinese today, conceptualizing the Qing as a colonial power would conflict with a number of largely unquestioned assumptions about China's past, and thus undermine historical constructs that lend legitimacy to current political practices in the People's Republic. Because twentieth-century China is accustomed to casting itself as a victim of colonial aggression in the nineteenth century, acknowledging its own past as a colonial power in the eighteenth century would deprive the modern nation of part of its self-definition. More threatening still, what would such an "admission" do to claims for territory conquered only under the expansionist Qing dynasty? Finally, claims to Chinese exceptionalism, which have been politically useful as a way of deflecting demands made by the Western world to conform to its standards regarding issues as diverse as human rights, copyright standards, and international trade agreements, would be undermined by admitting to a colonialist past similar to, and partially contemporary with, that of Europe.

Why has mainstream U.S. scholarship on China not engaged the question of China's involvement with the rest of the eighteenth-century world until quite recently? There may be several reasons. In the 1950s and 1960s English-language historiography on China was dominated by the work of John Fairbank, which centered on British and Chinese interactions in coastal areas of China during the mid–nineteenth century.[32] Out of the sources he used, and the period and geographic regions he examined, came a theory that China was changing (only) in response to contact, and confrontation, with the West. Fairbank's work did nothing to challenge Max Weber's, and even Karl Marx's, earlier theories portraying China as a static civilization with little internal impetus for change. While not himself locked into this model in later

32. John King Fairbank, *Trade and Diplomacy on the China Coast: The Opening of Treaty Ports, 1842–1854* (Cambridge, Mass.: Harvard University Press, 1948). "Response to the West" is the standard characterization of Fairbank's model for viewing China's entry into the modern world. See Paul A. Cohen, *Discovering History in China* (New York: Columbia University Press, 1984).

decades, Fairbank's early seminal work profoundly influenced the shape of the field.[33]

With improved access to Chinese archives in more recent decades, Western scholars working in Chinese history have been able to follow a more China-centered approach that actively rejects earlier assumptions of a stagnant, unchanging China. Proponents of this approach, however, in reaction to what became characterized as the "response to the West" paradigm, were often averse to exploring topics involving interaction between China and the Western world. A greater contribution could be made by taking an approach that would focus on diachronic change and/or regional diversity domestically.

It is now time for a new look at the way the Qing interacted with the eighteenth-century world, a look not colored by circumstances and events belonging to the nineteenth century, during which it experienced both extensive internal rebellions and repeated military defeats at the hands of more technologically advanced powers including both Britain and Japan. We need a new historiographic lens large enough to view the Qing dynasty both as an independent entity with its own history and momentum (as well as diversity and complexity), and as part of the larger early modern world. Only thus can we see how the Qing may have participated and even helped to shape trends in early modern world history, rather than only reacting to them.

The question of Qing connectedness with the early modern world is important for its own sake and to a better understanding of this period. But it also has much to contribute to the debate on the roots of nationalism in China. As hinted at already, Qing expansionism has implications for the way in which China views itself and is viewed today.

China's diverse population, its size, and/or its long history often take pride of place in the opening lines of current Chinese books on Chinese history, ethnicity, and geography aimed at the general reader.[34] A first sentence reading something like "China is a very large country with a remarkably long history that is populated by many different

33. For a succinct synopsis of Weber's influence on Western scholarship on China, see the introduction to William T. Rowe's, *Hankow: Commerce and Society in a Chinese City, 1796–1889* (Stanford, Calif.: Stanford University Press, 1984), 1–14.

34. See, for example, Dai Yi, ed., *Jian ming Qing shi* [A Concise History of the Qing] (Renmin chuban she, 1980), forward, 1; Ma Reheng and Ma Dazheng, eds., *Qingdai bianjiang kaifa yanjiu* [Research into the Development of Qing Dynasty Frontiers] (Zhongguo shehui kexue chubanshe, 1990), foreword, 1; Jiang Yingliang, ed., *Zhongguo minzu shi* [A History of China's Nationalities] (Minzu chubanshe, 1990), 1; *Miaozu jianshi* (Guiyang: Guizhou minzu chubanshe, 1985), publisher's note, 1.

nationalities [*minzu*]" is not unusual. Through frequent, almost mantra-like repetition, these attributes have become the defining features of the entity called China *(Zhongguo)*.[35]

Although the term *Zhongguo* is found in ancient Chinese texts, it came to be used consistently to refer to a unified China only relatively recently. For early periods the term is more appropriately translated as the "central states." The Ming (1368–1643) and Qing dynasties often referred to the territory they controlled as the Great Ming *(da Ming)* and Great Qing *(da Qing)*, respectively, as well as by a number of more literary terms—*Zhonghua* (Central [Cultural] Florescence), *Shenzhou* (the Spiritual Region), *Jiuzhou* (the Nine Regions), and *Zhongtu* (the Central Land)[36]—but not commonly as "China," *(Zhongguo)*. The present use of the term *Zhongguo* to refer to all of China is concurrent with the country's status as a modern nation-state.[37] Statements such as "China is a country with a very long history" tend to obscure this distinction in the interest of endowing the modern nation with the venerability accorded what is ancient.

As Benedict Anderson has argued, one of the recurring features of the modern nation-state is its relative youth coupled with a reliance on projecting the history of the nation back onto a geographic and cultural entity with a long past in order to assert a legitimacy born of the *longue durée*. The phenomenon of the modern nation claiming an age-old history has caught the attention of a number of scholars of nationalism, including Prasenjit Duara, who has looked specifically at its implications for the historiography of twentieth-century China.[38] The long history claimed by Chinese textbooks is not that of the nation (either the Republic of China—founded 1911—or the People's Republic of China—founded in 1949) but of a cultural heritage linked historically to a certain region. Absent, however, from the opening statement of books that invoke China's long history are what constituted this historical "China," who precisely is included in its cultural melting pot,

35. Cf., Peter Perdue, "A Frontier View of Chineseness" (paper presented at the Workshop on Renegotiating the Scope of Chinese Studies, Santa Barbara, California, March 13, 2000).

36. See Smith, "Mapping China's World," 53. Translations of these terms are Smith's.

37. The English translation of *Zhongguo* as "Middle Kingdom" may have had as much to do with nineteenth-century British and American perceptions (and representations) of China as ethnocentric as with the actual derivation of the term, but that topic would form the basis of a separate article.

38. Prasenjit Duara, *Rescuing History from the Nation: Questioning Narratives of Modern China* (Chicago: University of Chicago Press, 1995), especially pp. 27–33. See also Thongchai, *Siam Mapped*, 143–50.

why, and since when. The unquestioning reader is left to assume (incorrectly) that China today is what it always was—at least in terms of its geographical and ethnic composition.

In recent years much literature has been devoted to the formation of the nation-state and its roots in the early modern period. However, until Thongchai Winichakul's *Siam Mapped*, Asia was either neglected or seen only as fertile ground onto which a model first established in the Western world could be transplanted. Thongchai's work is so important because he deals with the historical process of transition from Siam as a kingdom to Thailand as a modern nation.[39] Without denying French and British historic interests in the area, he portrays Siam not as a victim but as a player that succeed rather well in defining its own boundaries and in subjugating, in the name of the modern nation-state, many now nameless kingdoms along its borders that once enjoyed a certain autonomy that was possible in a world not yet dominated by the logic of the modern nation-state. He shows that in competing successfully with Western colonialism Thailand took on the very values that first allowed colonial policy to succeed, and secondly contributed to the development of the logic of the system of nation-states.

But if the unquestioned value of nationalism has become key to perpetuating the military government in Thailand, so too certain present-day political imperatives have discouraged a consideration of seventeenth- and eighteenth-century Qing expansion as part of larger global trends. A shared history of struggle for territory using similar techniques of domination is threatening for several reasons. It invites recognition that China can justifiably be held to the same terms and standards as other world powers; "traditional" isolation from the world system would no longer be a viable justification for differences in legal and political practices to which human rights watchdogs object. More concretely it would pose a potential threat to Chinese national sovereignty and territorial integrity. If Tibet, Xinjiang, and other parts of modern China, let alone Taiwan, became part of what is now the People's Republic through Qing colonial conquest, would that not help to legitimize their claims to independence? Furthermore, envisioning the Qing as a colonial power integrated with the early modern world system would threaten the integrity of the model of a historic but nonetheless strangely atemporal China in which the dynastic system was

---

39. Siam officially changed its name to Thailand in 1941. The Thai term for Thailand translates literally as "the Thai nation." The name change, from "Muang Thai" ("the country of the Thai"), marked a conscious shift in the way in which Thai sovereignty was conceptualized (ibid., 49).

continuous and largely unchanged. The idea of a practically continuous history to which the PRC is a direct heir would be called into question.

No one would argue with the observation that today in China maps utilize to-scale, "Western" standards and techniques, and that ethnographic writing is viewed as politically sensitive. China shares the same concerns over possible nationalist separatist movements among its multi-ethnic populations that have plagued other postcolonial settings around the world. These shared conditions of modernity did not emerge fully formed only in the twentieth century, but have their roots in similar colonial pasts. Nor, while specifics surely vary from place to place, are they independent from the global context. To better understand their emergence we must be open to exploring contacts and intellectual exchange between China and the Western world during the early modern period.

Drawing parallels between the Qing dynasty and an expanding Europe poses a nagging question that cannot be ignored. While the Qing was undeniably an expansionist imperial power, is it really appropriate to refer to it as a colonial power as well? To address this question is no simple task. One first needs to explore what the term "colonial" actually means. Colonial systems are associated with territorial expansion, settlement, political dominance, economic exploitation, and often with a specific (if not well-defined) period of time—generally corresponding to the period of Western colonial expansion. More importantly, if more nebulously, "colonialism" has tended to be associated with attitude and place; like racism, it has come to be equated with continuing victimization caused by an evil perpetrated by "Western" countries and peoples against "non-Western" others during a period of European (and American) ascendancy.[40] Uncovering these emotionally charged associations helps us at least to understand what is at stake in speaking about the Qing as a colonial empire. It has as much to do with today's legacy as with what happened or did not happen in the past.

Part of what makes colonialism hard to nail down may be its relationship to nationalism, an equally difficult term to define. The reason for this is no coincidence; colonialism has in non-Western contexts often become nationalism's other—the threat against which the modern nation-state has defined itself. The assumptions around the terms "nationalism" and "colonialism" constitute precisely the reasons for their

40. See Frank Dikötter, *The Discourse of Race in Modern China* (London: C. Hurst, 1992), for an argument countering the assumption that racism can only be a product of the Western world.

significance.[41] The legitimacy of the governments of many modern nation-states is based at least partially on their role as protector of the people against colonial aggression. Uncovering, or admitting to, an indigenous colonial legacy might weaken the purity of such claims, for the modern nation-state cannot completely dissociate from its past.

I have chosen to use the term "colonial" in referring to the Qing state for a number of reasons. During the course of the seventeenth and eighteenth centuries, Guizhou province and other parts of the Qing empire were colonized by settlers who moved in to displace indigenous populations. As discussed further in chapter 4, the administrative governance of the southwest in particular was changed in order to bring the area under direct control of the central government for the first time. As territory was being colonized, "colonial" seems a reasonable adjective to use in describing the Qing.

But even more, I choose to use the term to highlight the similarities between the methods, technologies, and ideologies that the Qing employed in extending its geographical reach, and those used by European colonial powers during the same period.[42] As with European powers, both cartographic depiction of territory and ethnographic depictions of a given region's inhabitants were indispensable tools in the process of Qing empire building. Each in its own way attempted to represent, and thus to claim or define, the scope of an expanding China.

ORGANIZATION

Chapter 1 sketches out the ever-present delicate balance of power between the Manchu rulers and the majority Chinese population that they governed, and describes various techniques of Qing expansion. It also demonstrates evidence of the Kangxi emperor's knowledge of European expansion and his strategy for dealing with it. A description of the Qing *Imperial Illustrations of Tributaries (Huang Qing zhigong tu)* shows how the Qing conceived of its place in the world. This historically inspired conception was also in some ways adapted to changing times.

41. According to Thongchai, the intangible qualities of nationalism may be what make it so powerful. Our assumptions about the brotherhood of fellow nationals and about the "naturalness" of a nation's geographic borders help to form this powerful aura. See Thongchai, *Siam Mapped,* 3–6, 131–32.

42. For additional work on the Qing dynasty and comparative colonialism, see Di Cosmo "Qing Colonial Administration," and Perdue, "Comparing Empires" and "Boundaries, Maps, and Movement."

Chapter 2, "Mapping Territory," further develops the theme of the self-conscious Qing construction of its place in the world, fleshing out the argument that certain kinds of cartography sponsored by the Qing court attest to its involvement in, as opposed to isolation from, the early modern world. It begins with an introduction to the current state of scholarship on Chinese mapping and a proposal for a new direction in research on cartography in Qing China. The main body of the chapter examines informal international contacts and the exchange of geographical knowledge between European expatriates and Chinese scholars prior to the Kangxi emperor's reign; compares the chronological development of early modern mapping activity in France, Qing China, and Muscovy; and describes the Kangxi surveys of the Qing empire in which both Europeans and Manchus worked together to carry out astronomical surveys under the sponsorship of both the Qing and French courts. The concluding section discusses cartography as a constitutive component of early modern empires.

Chapter 3 turns the reader's gaze from mapping territory to "Depicting Peoples." It describes observations made in European contexts about the development of ethnographic writing and raises questions about their applicability to the Qing context. Both world areas have long histories of descriptive writings of "other" peoples that are briefly summarized. Although the development of descriptive writings of other peoples (like mapmaking) is not linear, yet trends can be identified, including movement toward what can be termed "early modern"—as defined by an emphasis on direct observation and empirical knowledge. The final portion of the chapter discusses ethnography's potential for political and colonial purposes.

The last four chapters of the book focus on the development of ethnographic depictions of non-Han peoples in southwest China, primarily Guizhou province, during the Qing dynasty. The first of these, chapter 4, introduces the geographical locus of this case study. It describes Guizhou's topography, peoples, historic attitudes regarding the education and acculturation of non-Han residents, the system of administration prior to the eighteenth century, and measures for governance instituted during the first half of the eighteenth century. The latter included mapping Guizhou's territory, instituting a new, more centralized system of administration in much of the province, and attempts at formalizing policies of assimilation under the Yongzheng emperor (r. 1723–1735).

Chapter 5 discusses the development of ethnographic writing in the province based on a series of local gazetteers and histories dating from as early as 1560 to as late as 1834. These texts are crucial to our understanding of the development of ethnographic writing in the context of Qing frontier expansion. Unlike other ethnographic documents described in chapters 6 and 7, they are precisely dated, making it possible to detect and record change in content (and priorities) over time. In each case we know who the authors were, what official positions they held, and hence what their formal interest in the province was. Over this period of more than 250 years we see marked growth in the geographical scope of ethnographic inquiry, the emergence of a complex taxonomy of non-Han groups, expanded evidence for direct observation as a basis for ethnographic depiction, and an increase in the recording of successful instances of pacification or acculturation.

Chapter 6 looks at the emergence of the Miao album as a distinct genre during the eighteenth century. The albums are described in detail, including how they differ from the local gazetteers and histories examined in the previous chapter. Information recorded in half a dozen extant prefaces provides clues as to their provenance, the occupations of their authors, and their uses.

Chapter 7 examines later developments in the Miao album genre, including their changing nature and uses over time and the scholarship that they have generated.

CHAPTER ONE

The Qing Empire: Constructing a Place in the Eighteenth-Century World

During the century spanning 1660–1760, the Qing empire doubled the amount of territory it controlled. Its enlarged scope, with the exception of present-day Mongolia, largely forms the shape of the territorial claims of the People's Republic of China today (see map 1). This expansion did not take place within a vacuum, but in the context of global population growth[1] and early modern colonial exploration and expansion worldwide. Not only did the Qing empire interact with early modern forces affecting the whole globe during this period, but the results of its imperial expansion have shaped the scope and nature of the modern Chinese nation-state in concrete ways. Many non-Han, or culturally non-Chinese, peoples who were incorporated into the empire during the Qing now help to make up the fifty-five officially recognized "minority nationalities" of the People's Republic.

The Qing was distinctive as a nonnative dynasty; the ruling family was of Manchu descent.[2] That China formed only one part of a distinctively Qing empire is graphically apparent from the 1719 scroll version of the Kangxi atlas in which place names within China proper appear in Chinese characters, and all other areas of the empire are labeled in Man-

---

1. During the same period China's population is estimated to have at least doubled. See Ho Ping-ti, *Studies on the Population of China, 1368–1953.* (Cambridge, Mass.: Harvard University Press, 1959), 278, 281–82; William Lavely and R. Bin Wong, "Revising the Malthusian Narrative: The Comparative Study of Population Dynamics in Late Imperial China," *Journal of Asian Studies* 57.3 (August 1998): 714–48.

2. For a recent and extensive study on the Qing Imperial Household, see Evelyn S. Rawski, *The Last Emperors: A Social History of Qing Imperial Institutions* (Berkeley, Los Angeles, and London: University of California Press, 1998).

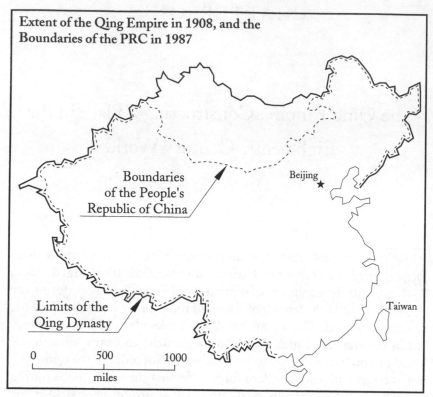

Map 1 The Extent of the Qing Empire in 1908 and Boundaries of the People's Republic of China in 1987. Adapted from *The Historical Atlas of China (Zhongguo lishi ditu ji)*, vol. 8, Tan Qixiang, ed. Shanghai: Cartographic Publishing House, 1987. Cartography Laboratory, Department of Anthropology, University of Illinois at Chicago.

chu.[3] The Manchus had their own (written and spoken) language, a nomadic tradition, and customs and practices that were different from their Chinese neighbors. In order to successfully claim the "Mandate of Heaven," which according to Chinese cosmology legitimized the emperor's rule, the new rulers had to successfully meet a number of challenges. Not least was to gain a thorough familiarity with the classical Chinese language, the dictates of Chinese custom, and particularly the values and practices prized by the elite in order to convince this crucial segment of the population that, even as foreigners, they could legitimately fill the requirements demanded of the imperial house. At the

3. Matteo Ripa, "A Map of China and the Surrounding Lands Based on the Jesuit Survey of 1708–1716" (British Library, 1719), K.top.116.15, 15a, and 15b. The three scrolls are 79″ tall by 125″ long, 60.5″ tall by 109.5″ long, and 38″ tall by 93.5″ long, respectively.

same time they strove to protect their own cultural identity. Most court documents were recorded in both classical Chinese and Manchu;[4] intermarriage was not allowed; and Manchu women were forbidden to bind their feet. Manchu garrisons were segregated from the rest of the population.

For at least a century and a half the dynasty was remarkably successful in its efforts to build and expand the empire. Its achievements can be attributed to the balance it maintained between the creative use of its cultural ties with Inner Asia to form alliances to the north and northwest, the fostering of an ideology of universal rule based on Confucian principles, and an awareness of and openness to early modern technologies.[5] Over time these three sets of demands would become increasingly difficult to juggle.

In Inner Asia the Manchus successfully manipulated common points of reference as a base for establishing legitimacy. The Manchus' religious practices, including Tibetan Buddhism, and their martial and nomadic traditions, opened various avenues for legitimizing the extension of Manchu rule. The Qing made use of both secular power and religion in strengthening its hold on Tibet and Mongolia.[6] The Qianlong emperor (r. 1736–1795) in particular exercised the political manipulation of religion by representing himself as a bodhisattva, thus fostering the adulation and compliance of adherents of Tibetan Buddhism.[7] Strategic marriages with rival powers were practiced in the north and northwest.

The south and southwest, populated largely by ethnically non-Han and culturally non-Chinese[8] peoples, presented special problems. Here the Manchu heritage provided no particular advantages or status to the imperial government. Intermarriage was hardly deemed appropriate with the indigenous population, nor did religion provide a common

4. Pamela Kyle Crossley and Evelyn S. Rawski, "A Profile of the Manchu Language in Ch'ing History," *Harvard Journal of Asiatic Studies* 53.1 (June 1993): 63–102.

5. Pamela Crossley describes the Qing cultivation of what she refers to as different "constituencies" in her *Translucent Mirror*.

6. See Chia Ning, "The Lifanyuan and the Inner Asian Rituals in the Early Qing (1644–1795)" *Late Imperial China* 14, no. 1 (June 1993): 60–92, and "The Manchu Rule in Early Modern Mongolia: Religious or Secular?" presented at the annual meeting of the Association for Asian Studies, March 26–29, 1998; Waley-Cohen, "Religion, War, and Empire-Building"; and James Hevia, "Lamas, Emperors, and Rituals: Political Implications of Qing Imperial Ceremonies," *Journal of the International Association of Buddhist Studies* 16.2 (1993): 243–78.

7. David M. Farquhar, "Emperor as Bodhisattva in the Governance of the Ch'ing Empire," *Harvard Journal of Asiatic Studies* 38 (1978): 5–34.

8. By culturally non-Chinese I mean that they did not participate in the dominant culture as defined by the broadest of Confucian norms and values.

bond. Thus other means and models of incorporating territory and peoples in these regions had to be found.

Initially the Qing rulers relied largely on an established Chinese model of bureaucratization and Confucianization for dealing with minority groups in these frontier areas. A system of local administration in which indigenous leaders *(tusi)* ruled over non-Han populations continued in use during the Qing.[9] At the same time, in areas with relatively more contact between native and Han populations, the local (largely Han) administration aspired to an ideal of sinicization[10] through education as a means to bring the various non-Han border groups into sharing a system of cultural values with their Han neighbors.[11] However, this inherited dynastic model of dominance via the sinicization of non-Han groups through education came to conflict with the evolving Manchu vision of empire. The Qianlong emperor in particular consciously struggled to preserve (and in some ways create) Manchu racial and cultural identity.[12] How then could the court justifiably promote the sinicization of other non-Han groups?

Two separate but not necessarily mutually exclusive paths promised a way out of this conundrum. One path was to look beyond previous precedents of sinicization entirely by turning to early modern technologies of rule not rooted narrowly in Confucian thought. In the southwest, for example, resources were aimed at the production of knowledge about frontier areas and the peoples who inhabited them. Accurate, detailed maps and ethnographic accounts became key to achieving and maintaining control. Efforts in these areas reflect the kinds of epistemology and technologies of representation typical of the early modern period in

9. See chapter 4 for more on the *tusi* system.

10. A two-way model of acculturation is more appropriate to the situation than actual sinicization, but I use the term here because it reflects the rhetoric used by officialdom to express these goals *(hanhua,* to transform into Han). For a discussion of the problems accompanying the use of the term "sinicization," see Pamela Crossley, *Orphan Warriors: Three Manchu Generations and the End of the Qing World* (Princeton: Princeton University Press, 1990), 223–28. For ongoing debate on this topic, see Evelyn S. Rawski, "Reenvisioning the Qing: The Significance of the Qing Period in Chinese History," *Journal of Asian Studies* 55.4 (November 1996): 829–50; and Ping-ti Ho, "In Defense of Sinicization: A Rebuttal of Evelyn Rawski's 'Reenvisioning the Qing,'" *Journal of Asian Studies* 57.1 (February 1998): 123–55.

11. See Claudine Lombard-Salmon, *Un exemple d'acculturation chinoise: La province du Guizhou au XVIIIᵉ siècle* (Paris: École Française d'Extrême-Orient, 1972); and William T. Rowe, "Education and Empire in Southwest China: Ch'en Hung-mou in Yunnan, 1733–1738," in *Education and Society in Late Imperial China, 1600–1900,* ed. Benjamin A. Elman and Alexander Woodside (Berkeley, Los Angeles, and London: University of California Press, 1994), 417–57.

12. Pamela Kyle Crossley, "*Manzhou yuanliu kao* and the Formalization of the Manchu Heritage," *Journal of Asian Studies* 46.4 (November 1987): 779–81. See also her *Translucent Mirror*.

Europe. Similar processes in both world areas, which led to recognition of the finite nature of the globe and the expansion of other world powers, encouraged similar responses.

The other strategy was to create a universal ideology of rule based on, but extending beyond, Confucian principles. The goal would be to convince not only the ethnic Han population but also other peoples in the expanding empire that there was a place for everyone under the umbrella of Qing rule. Confucian rituals involving the emperor's position as Son of Heaven and long-established idioms of the emperor as the figurative center of the empire were fundamental to this approach.

CONNECTIONS TO THE EARLY MODERN WORLD

The conflicting demands, then, of leading a multiethnic empire posed a dilemma to the court. How could they both perpetuate (or indeed create) the Manchu identity crucial to controlling Inner Asia, and simultaneously foster the Confucian ideology necessary to gain Chinese loyalties? The conundrum may have induced the Qing court to experiment with the first tactic mentioned above: employing technologies (and technicians) that were neither identifiably Chinese nor Manchu to assist in the process of empire building. Jesuit priests and lay missionaries, who adopted Confucian dress and learned both Manchu and Chinese in order to adapt themselves culturally to their host country, worked in extremely sensitive areas where they held positions of authority. They headed the imperial board of astronomy, conducted cartographic surveys of the empire, planned weapons manufacture (primarily cannons), worked as tutors and advisers to several emperors, depicted the court and the empire in imperially commissioned paintings, and no doubt served as conduits of information about the outside world as well. During the early years of the dynasty their status as outsiders may have made them less threatening to the Manchus than relying only on the assistance of Han officials, who might have had more reason for wanting to subvert the alien dynasty. As long as their loyalty could be assured, their relationship to the Manchu court was less complex than that of Han Chinese bureaucrats. More importantly, the emperor seems to have recognized the potential benefits that new technologies from outside offered.

Tutored as a youth by Adam Schall, the Kangxi emperor (r. 1662–1722) was well versed in math and science as taught in the West, in addition to having the best possible Chinese and Manchu education. His patronage of Chinese scholarship and continuing support of Manchu tradition is well recognized, but he also retained an active interest in Western learning as an adult:

> In 1691 he ordered European scholars to translate a core of philosophy into Manchu . . . and in order to satisfy his desires and his curiosity he also had translated into Manchu the most interesting and newest discoveries recorded in the *Mémoires* of the Academy of Sciences of Paris and numerous other European publications.[13]

His interest in science as practiced in Europe was deeper than mere curiosity.[14]

As set forth above (see introduction), there are a number of complex and interlocking historiographical reasons why Kangxi and his immediate successors' appreciation of "Western" technology has not generally been recognized as more than a passing interest in foreign curiosities.[15] Another reason is that accumulating hard, written evidence of the emperor's knowledge of and engagement with Western learning is not a simple task. The historical record in Chinese is unlikely to record such involvement or fascination because for the vast majority of readers of classical Chinese the imperial authority derived from the emperor's mastery of Chinese, and more specifically Confucian, high culture. Serious interest in competing epistemological systems is not something the Chinese literati would have been likely to condone, nor would the Kangxi emperor have cared to flaunt it in certain contexts. French documents written by eyewitnesses at the court do, however, provide a window onto some areas of his interaction with technologies and systems of thought emanating from Europe. A careful reading of the letters of Antoine Gaubil, who served at the Qing court from 1722 till his death in 1759, and of Joseph-Anne-Marie de Moyriac de Mailla's *Histoire générale de la Chine,* based on a translation of a Manchu-language history of China, provide evidence that the Kangxi emperor had a good grasp of, and valued, the information and technology available through channels that had been established with Europe.

---

13. De Mailla, *Histoire générale de la Chine,* vol. 11, 364.

14. On the Kangxi Emperor, see Lawrence D. Kessler, *Kang-Hsi and the Consolidation of Ch'ing Rule, 1661–1684* (Chicago and London: University of Chicago Press, 1976), especially pages 146–54; and Jonathan D. Spence, *Emperor of China, Self-Portrait of K'ang-Hsi* (New York: Vintage Books, 1975).

15. Waley-Cohen's 1993 article "China and Western Technology in the Late Eighteenth Century" is the most notable exception.

De Mailla's *Histoire* expresses the opinion that Kangxi's success as a ruler was due in part to his openness:

> Sovereign of two great nations of such opposing characters, he had become their leader in the sciences and in the practices which they esteemed most highly. Kangxi soon learned, from the Europeans attached to his court, to what degree the sciences and arts had been pushed toward perfection in the West. He had too great an admiration [goût] for this to limit himself to Chinese books. To the astonishment of the Chinese he traced a new route, one that a misplaced presumption and overattachment to their ancient practices had kept them from following.[16]

While we can detect a certain amount of self-congratulatory Eurocentrism in the tone of the above comment, it does not invalidate the observation itself.

The court's interest in Jesuit knowledge and skills shows sensitivity to both global concerns and the importance of early modern technologies. Eighteenth-century Qing expansionism should not be seen as operating in a regional vacuum. The Kangxi emperor was mindful of expanding European (and Russian) mercantile, and increasingly colonial, interests in central, south, southeast, and east Asia. He recognized that the Qing would need to stake out its own territory internationally against competing claims of other empires.

Antoine Gaubil recorded his memories of a meeting at court soon after Kangxi's death at which the late emperor's reflections on Europe and the politics of empire were discussed. The memoir reflects Kangxi's sophisticated understanding of the difference between the Russians, Dutch, and Portuguese and the potential threat that each posed to the Qing in the future. On the Russians:

> The Russians used only to form a small state in Europe, but they have become powerful there and nearly destroyed the bellicose nation of Sweden. They have extended themselves toward the East, and they have ports and ships in Europe, Cachan, Astrakhan, on the Caspian Sea, and in Siberia. They have large armies, and many different peoples pay tribute to them. . . . They have allied themselves with our enemies, the Eleuths. The Russians have become our neighbors bordering several rivers (Selinghe, Toula, Kerlons, Sahalien Oula) and there are Central Asians (Tartars) in these regions who pay tribute to them, as well as forts, towns, and troops equipped with good weapons. . . . They are thinking of constructing forts and ships which would allow them to come into the sea around Japan, Korea, and also China's maritime provinces.[17]

16. De Mailla, *Histoire générale de la Chine*, vol. 11, 361–62.

17. Antoine Gaubil, S.J., *Correspondance de Pékin, 1722–1759* (Geneva: Librairie Droz, 1970), 710 (letter to P. Berthier, 1752).

On the Dutch:

> The Dutch are very powerful in the seas of the Indies. They have destroyed
> the Portuguese [in the area], some remnants of which are now in Macao.
> There are many Chinese among them. We were able to chase them from
> Taiwan, but they are very powerful in Sumatra, Java, Malacca, Borneo, the
> Moluccas, and other places near the Chinese province of Yunnan. They have
> good soldiers, an infinite number of good ships, and a lot of money.[18]

On the Portuguese:

> The Portuguese of Luzon have many Chinese and they could easily become
> very powerful in the countries neighboring China and Japan. Their king in
> Europe is extremely rich, and possesses great states far from Europe. He is of
> the same family as the King of France, a powerful and bellicose nation which
> is esteemed and respected on land and sea throughout the world.[19]

These remarks show evidence of a sophisticated grasp of both China's
place in world geography and of international affairs, including European expansionism.

The Kangxi emperor's observations went even further, weighing a
potential future threat posed by European expansionism:

> The Russians, Dutch, and Portuguese, like the other Europeans, are able to
> accomplish whatever they undertake, no matter how difficult. They are in-
> trepid, clever, and know how to profit themselves. As long as I reign there is
> nothing to worry about from them for China. I treat them well, and they like
> me, respect me, and attempt to please me. The kings of France and Portugal
> take care to send me good subjects who are clever in the sciences and arts, and
> who serve our dynasty well. But, if our government were to become weak, if
> we were to weaken our vigilance over the Chinese in the southern provinces
> and over the large number of boats that leave every year for Luzon, Batavia,
> Japan, and other countries, or if divisions were to erupt among us Manchus
> and the various princes of my family, if our enemies the Eleuths were to suc-
> ceed in allying with the Tartars of Kokonor, as well as our Kalmuk and Mon-
> gol tributaries, what would become of our empire? With the Russians to the
> north, the Portuguese from Luzon to the east, the Dutch to the south, [they]
> would be able to do with China whatever they liked.[20]

From Gaubil's letter one obtains a self-image of the Qing emperor as a
player on the international scene. He thought of himself as on a par with
the leaders of a healthy and expansionist Europe. He participated with

18. Ibid., 710.
19. Ibid., 710–11.
20. Ibid., 711.

them in patronage of the arts and sciences, taking part in the kinds of activities that kept them strong.

Chinese sources are not completely silent on the topic of the Kangxi emperor's knowledge of the Western world. The Veritable Records of the Qing Dynasty record a number of significant statements by the Kangxi emperor. The occasion of a visit from a Russian ambassador in 1693 prompted the following remark:

> Although it is splendid that the foreign vassal should come to present tribute, we fear that after many generations, Russia might cause trouble. In short, as long as the Middle Kingdom is at peace and is strong, foreign disturbances will not arise. Therefore, building up our strength is a matter of fundamental importance.[21]

An edict written in 1716, concerned primarily with various aspects of the naval defense of the empire, echoes similar concerns: "After hundreds of years, We are afraid that [our country] will suffer injury from the overseas countries, for example, from the European countries. This is only a prediction."[22] While the Kangxi emperor was confident of Qing strength in his day, he was well aware of the dynamic and potentially threatening nature of the countries with which his empire was experiencing increasing contact.

As we will see in chapter 2, one of the means of competing was to stake out Qing territory internationally against claims of other empires. This was achieved by expanding on the ground militarily and on paper cartographically using the emergent international discourse of early modern mapping, which relied on to-scale representations based on information collected by means of astronomical surveying techniques.

QING IMPERIAL ILLUSTRATIONS OF TRIBUTARIES: A CLAIM TO UNIVERSALITY

While knowledge of current events and keeping up with new technologies abroad was crucial to Qing competition for international prominence, these were not sufficient means to foster confidence in Qing rule domestically. Other ways rooted in methods that resonated with Chinese tradition had to be explored as well. To this end the Qing

21. *QSLSZ*, 160: 26–27 (November 24, 1693). Translation is from Lo-shu Fu, *A Documentary Chronicle of Sino-Western Relations, 1644–1820*, Association for Asian Studies: Monographs and Papers, no. 22 (Tucson: University of Arizona Press, 1966), 106.

22. *QSLSZ*, 270: 14–15. Translation is from Lo-shu Fu, *Documentary Chronicle of Sino-Western Relations*, 123.

emperors fostered a type of universalist Confucianism. Evidence of the attempt to create a philosophy of rule that would benefit subjects of the empire—regardless of their cultural origin—can be seen in the Qing Imperial Illustrations of Tributaries *(Huang Qing zhigong tu)* commissioned by the Qianlong emperor in 1751. The idea behind the work was to portray all the different peoples who paid tribute to the Qing empire in one place and in a uniform format.

Several different versions of the Qing Imperial Illustrations of Tributaries exist.[23] The Xie Sui version is the most elaborate and clearly an imperial showpiece. The work consists of four hand scrolls, each approximately thirteen and one half inches tall. Accompanying text, in both classical Chinese and Manchu, describes the location of the peoples depicted and something of their customs, livelihood, costume, and religion. Each illustration, executed in color on silk, includes both a man and a woman. Much attention is given to details of costume. Although no background scenery or landscape is included, one or both of the figures may hold some sort of artifact, such as a tool, a weapon, or a container of some sort. A figure might also appear on horseback, or seated at a loom. (See plates 2 and 3.) The woodblock versions, although less elaborate and only in Chinese, are otherwise quite similar. Lists of material that were paid in tribute were not included. This kind of detailed accounting was recorded separately in another location. The focus here was on the peoples.

The geographical scope of the Qing Illustrations of Tributaries was extremely broad, ranging from non-Han groups that came to be under the direct control of the Qing government to representative figures from countries that would have considered themselves totally independent— more nearly trading partners. Their order of appearance follows a systematic progression from outer to inner and east to west. The first chapter *(juan)* contains peoples from foreign countries *(guo)* beginning to the east with Korea, then progressing south toward the Ryukyu

---

23. I have seen copies of three different editions of the Illustrations of Tributaries: a woodblock version dating from the twenty-sixth year of the Qianlong emperor's reign (1761), housed in the Fu Ssu-nien Library at Academia Sinica in Taiwan; the "Xie Sui" version, named for its compiler, which is undated but probably completed in stages sometime between 1761 and 1775 (reprinted in Chuang Chi-fa, ed., *Xie Sui "zhigong tu" manwen tushuo jiaozhu* [Taipei: guoli gugong bowuyuan, 1989]); and the *Siku quanshu* edition (prefaced 1777 although not published until 1790). Another edition, which I have not seen, was published in 1805 (see Chuang, *Xie Sui,* 1). For a description of the three versions and their fundamental differences, see Laura Hostetler, "Chinese Ethnography in the Eighteenth Century: Miao Albums of Guizhou Province" (Ph.D. diss., University of Pennsylvania, 1995), 185–86.

islands, Annam, Thailand, the Sulu Archipelago, Burma, and "Nan zhang"—a country located to the northeast of Thailand, west of Annam and bordering on Yunnan. Next, representatives of larger European countries are shown in the following order: Swiss, Hungarian, Polish, black slaves of Europeans, and members of religious orders. These were followed by peoples of smaller European countries: English, French, and Swedish. The entries then continued with figures representing Japan, Holland, several countries whose names I cannot identify, Luzon, Java, Malacca, and the Solomon Islands.

The second and third chapters contain various *fan ren*—peoples not recognized as subjects of specific countries *(guo)*, but also not part of the regular administrative system of the Qing empire. They include, for example, peoples from Tibet, various Islamic peoples (labeled with the suffix *hui min)*, and others. The remaining entries, which constitute over two-thirds of the Illustrations of Tributaries, are of non-Han peoples located within the Chinese provinces of Fujian, Hunan, Guangdong, Guangxi, Gansu, Sichuan, Yunnan, and Guizhou. The headings for these groups always list the administrative jurisdiction, often county, in which the peoples portrayed lived.

The Qing "tribute system" has often been taken as exemplary of China's ethnocentrism, sense of (false) superiority, or even isolationism. For some it is a symbol for the country's backwardness in refusing to understand and interact with the modern world on modern terms. However, as pointed out by a number of scholars working in this area, the concept of tribute during the Qing is not well understood; no single "system" *(zhidu)* as such existed.[24] Tribute was both internal, submitted from within China's provinces, and external, coming from farther afield. Furthermore, different expectations were applied to different regions. Finally, we need to recognize that tribute could and did have different meanings at different times and in different contexts.[25]

24. For recent work on Qing tribute, see the following: John E. Wills, Jr., "Tribute, Defensiveness, and Dependency: Uses and Limits of Some Basic Ideas about Mid-Qing Dynasty Foreign Relations," *American Neptune* 48 (1988): 225–29; Hevia, *Cherishing Men from Afar.* A 1995 panel at the annual meeting of the American Historical Association on "Tribute, Trade, and Imperial Power in China under the Qing Dynasty, 1644–1911" included the following papers on the topic: Nancy Park, "Imperial Tribute and the Rise of Official Corruption in Mid-Qing China, 1735–1796"; Chia Ning, "Mongol Tribute in Seventeenth- and Eighteenth-Century China: The Significance of Gift Exchange in the Formation of National Boundaries"; and James Millward, "Xinjiang and the Trouble with Tribute."

25. For differences in Qing administration of frontier areas from previous dynasties, see Di Cosmo, "Qing Colonial Administration."

Illustrations of Tributaries *(zhigong tu)* were not a new genre in China during the Qing. During the Liang dynasty (502–557), several efforts were made to picture peoples of other states. Emperor Wu (520–549) is said to have commissioned an artist to make a book entitled *Fangguo shitu* (Illustrated Descriptions of Envoys from All Lands) that included representatives of twenty different states.[26] The first known instance of tribute illustrations, or *zhigong tu*, was also commissioned under the Liang, by emperor Yuan (r. 552–554). It contained representatives of thirty different states.[27]

Foreign peoples often appeared in the art of the Tang dynasty (618–906). The court artist Yan Liben (7th c.) is famous for his Illustrations from Western Border Regions. The Tang dynasty also gave rise to *Wanghui tu*, or Illustrations of Meetings with Kings, that later Miao album prefaces sometimes invoked as a precedent. These were paintings commissioned by the emperor of China to record visits of foreign kings or dignitaries to his court. Illustrations of Tributaries were also painted during the Song dynasty (N. Song 960–1126; S. Song 1127–1280).[28] Although there are a number of examples of earlier *zhigong tu*, we cannot automatically equate them with the later Qing work. Certainly the Qing emperors were aware of, and even derived a degree of authority from these precedents, but the format and purposes of the Qing Illustrations of Tributaries were not identical to those of earlier periods.

In a discussion of Qing ritual, James Hevia argues that we have too often done Chinese history a disservice by viewing China as possessing "culture" in contrast to our treatment of Europe as having history, or cultural forms whose meaning and practice have changed over time. In his work on the Macartney embassy of 1793, his aim is to "cease interpreting the Macartney embassy as an encounter between civilizations or cultures, but as one between two imperial formations, each with universalistic pretensions and complex metaphysical systems to buttress such

---

26. F. Jaeger, "Über chinesische Miaotse-Albums," *Ostasiatische Zeitschrift* (Berlin) 5 (1917): 82. See also Chuang Chi-fa, "Xie Sui zhigong tu yanjiu" [A Study of the *Tribute Presenting Scroll* by Hsieh Sui], Proceedings of the 1991 Taipei Art History Conference, National Palace Museum, Taipei, 1992, 773.

27. F. Jaeger, "Über chinesische Miaotse-Albums," *Ostasiatische Zeitschrift* (Berlin) 4 (1916): 266; and Friedrich Hirth, "Über die chinesischen Quellen zur Kenntniss Centralasiens unter der Herrschaft der Sassaniden etwa in der Zeit 500 bis 650," *Wiener Zeitschrift für die Kunde des Morgenlandes* 10 (1896): 225–41, 227.

28. Jaeger, "Über chinesische Miaotse-Albums" (1916), 267.

claims."[29] By historicizing both British and Chinese concerns and behaviors, he reaches beyond the assumptions that have limited our view of this encounter in the past, opening up new horizons for inquiry.

We can profit from approaching the Qing Imperial Illustrations of Tributaries in a similar critical spirit. Looking at the Illustrations of Tributaries in the context of the early modern world, we can see that the methodology behind it, its claims, and even its format are not out of step with those of other contemporaneous imperial powers. Part of what made the Qing Imperial Illustrations of Tributaries different from its precedents was the method of its compilation. In 1750 the governor-general of Sichuan, Celeng, received an imperial edict ordering him to make annotated illustrations of the various non-Han groups under his jurisdiction:

> On the eleventh day of the eighth month of the fifteenth year of the reign of the Qianlong Emperor, I [Celeng] received an imperial edict written and sent by the Grand Secretary, Loyal Prince, Fuheng, ordering me (your servant): "Take the western barbarians *(xi fan)*,[30] and the Luoluo with which you are familiar, and make illustrations and commentary concerning the appearance of the men and women, their dress, ornamentation, clothing, and customs. As for those with which you are not familiar, you do not need to send [anyone] to make inquiries. . . ." Whereupon I respectfully [took] those *yi* regions *(yi di)* I had experience of and those barbarian subjects *(fan min)* I had officially met *(jiejian)*, and also consulted with the civil and military [officials] in those jurisdictions, and had twenty-four illustrations made. I also followed each [illustration] with a clear explanation of the land, customs, costumes, preferences, and general circumstances of the places. Upon completion they were wrapped and respectfully submitted for imperial inspection.[31]

The 1750 edict was especially precise in stating that information submitted to the throne should be based on direct experience. Only those groups with which the governor-general was personally familiar were to be reported on in this preliminary survey. Celeng himself uses the words "experience" *(jingli)*, and "meet" *(jiejian)* in reporting on how he compiled the illustrations and texts he submitted. Information not based on his own observation he gained from consultation with the civil and military officials in charge of those areas in question. Presumably not only

---

29. Hevia, *Cherishing Men from Afar*, 25.

30. This refers to "barbarians" to the west of China, not to what we think of now as "Westerners."

31. *Secret Palace Memorials of the Ch'ien Lung Period* (Taipei: National Palace Museum, 1982). Memorial dated the seventeenth day of the eleventh month of the sixteenth year of the Qianlong emperor's reign (1751).

the illustrations, but also the textual descriptions about the "land, customs, preferences, and general circumstances of the appropriate places," were also based on these observations and personal consultations.

One year later, in 1751, Celeng received an edict in which the emperor stated his vision of the project and gave special instructions for carrying it out.

> My dynasty has united the vast expanses. Of all the inner and outer barbarians *(nei wai miao yi)* belonging under its jurisdiction, there are none that have not sincerely turned toward Us and been transformed. As for their clothing, caps, appearance, and bearing, each [group] has its differences [from the other groups]. Now although we have likenesses *(tuxiang)* from several places, they are not yet uniform and complete. Gather together the several models *(tushi)* that we now have, and deliver them to each of the governors and governors-general near the borders *(bian)*. Order them to take the Miao, Yao, Li, and Zhuang under their jurisdiction, as well as the various outer barbarians *(wai yi fan zhong)*, and according to these examples copy their appearance, bearing, clothing and ornaments, make illustrations and send them to the Grand Council for classification and arrangement for presentation and inspection [by the emperor].[32]

The emperor was aware of the wide-ranging diversity of his realm, and desired to record it—but in a standardized format. His desire for likenesses that were both "uniform and complete" is reminiscent of the Kangxi emperor's desire to have one atlas "which would unite all the parts of his empire in one glance."[33] Because of his desire for consistency, the Qianlong emperor actually had prototypes made and sent to the provinces and border areas for the painters to follow.

The highly political nature of the mission is not something that is only apparent to a late twentieth-century onlooker. The emperor himself was keenly aware of the need for secrecy and subtlety in carrying out the project. The governors who received the edict requesting submission of annotated illustrations were directed to collect information only in conjunction with other business that would require interactions with the target peoples. Officials were not to call attention to the project. Literally they were warned: "Keep disclosure of this to a minimum; [otherwise] it may perhaps give rise to suspicion and dread."[34] The Qianlong emperor made it clear that the collection of information was to be based on direct observation, but also that it was to be carried out with secrecy from the people it was to represent.

32. *QSLGZ*, 390: 8–9.
33. De Mailla, *Histoire générale de la Chine*, vol. 11, 314.
34. *QSLGZ*, 390: 9.

Figure 10 Detail from a Seventeenth-Century World Map. Claes Janszoon Visscher, Amsterdam, c. 1617. According to Rodney W. Shirley's *The Mapping of the World*, "[a] broad panel at the top centres on the figure of Europe receiving homage from natives of other continents. The superior role of Europe, civilised and in fashionable dress, contrasts with scenes of the idolatry and cannibalism in other parts of the world" (317). Österreichische National Bibliothek.

The Qing Imperial Illustrations of Tributaries was a global "cultural map" that portrayed the Qing dynasty as a center of cultural imperialism (seen by its makers in a positive sense). Its implicit claims parallel those of the cartouches found on many European world maps from the same period that show Europe as the personified center of the world receiving tribute from all quarters. Another, more extensive, European work reveals an explicit effort at cataloging other peoples the world over. Jacques Grasset-Saint-Sauveur's 1798 *Tableaux des principales peuples de l'Europe, de l'Asie, de l'Afrique, de l'Amérique, et les découvertes des capitaines Cook, La Pérouse, etc. etc.* contains a significant amount of ethnographic commentary. The book is accompanied by a foldout chart listing the different peoples of the world, and also five color engravings depicting the peoples of Asia, Africa, Europe, America, and the peoples inhabiting the islands most recently discovered by Captain Cook (plate 6). Like these European images (see figures 10 and 11, and plates 4 and 5), it was not created only as an informational document, but also contained persuasive or idealized elements in the view of the world that it propagated. It drew its strength and legitimacy from Confucian concepts sanctioned by China's history, even while incorporating new meanings. Viewing it as a historical document we cannot take all the information it contains at face value (for example many of the countries included would have contested their status as tributaries), but it does provide a record of Qing attempts to make sense of the eighteenth-century world and of its claims as a competing center. Viewed from this vantage point the Qing Imperial Illustrations of Tributaries is no longer a relic of "tradition," but becomes a viable representation of an albeit idealized conception of the world at the point at which it was made, and of the Qing dynasty's role therein.

The Illustrations of Tributaries can be seen from another vantage

Opus, nunc denuò ab ipso Auctore recognitum, multisque locis castigatum, & quamplurimis
nouis Tabulis atque Commentarijs auctum.

Figure 11 Frontispiece to the First Edition of Abraham Ortelius's *Theatrum Orbis Terrarum,* Antwerp, 1579. Note the figure of "Europa" at the top. In the edition housed at the Universiteitsbibliotheek, Amsterdam, the figures are labeled Europe, Asia, Africa, and America. (See Wolter and Grim, *Images of the World,* 74.) Courtesy The Newberry Library, Chicago.

point as well. Its contents provide evidence that by 1750 the court had come to pursue a policy that championed a political unity made up of the many faces of cultural diversity. The work not only proclaimed the Qing's centrality, but also reflected and recorded the variety of ethnic and cultural diversity within and beyond the empire. Bringing this heterogeneous assortment of peoples and customs together for display in one consistent format may in itself have been a symbolic statement about the creation of the Qing empire,[35] and about Qing China's place in the world. The project of bringing together diversity within a uniform format reflected the emperor's vision of the Qing. Like the empire that it represented in microcosm, the Illustrations of Tributaries was made up of a multiplicity of peoples. Yet just as the diversity displayed in the illustrations was brought together within a greater unity of form, the various peoples of the empire (and even beyond its borders) were united in their recognition of (if not subservience to) the Son of Heaven. Today many of the peoples included in the Illustrations of Tributaries are still held together (some of them against their will) as part of the modern Chinese nation, albeit with different ideological glue. Its role in the creation of the currently accepted ethnic composition of the modern Chinese nation-state needs to be recognized as one of the Qing dynasty's major achievements—and as consistent with trends of empire during the early modern period internationally.

35. Han members of the Confucian bureaucracy may have understood the project differently from Manchus. The former may have stressed goals of unification over the diversity represented, whereas the latter may have more nearly recognized diversity as an essential component of unity.

CHAPTER TWO

Mapping Territory

O ne of the most graphic ways of seeing the Qing construction of its place in the world is through its use of maps. Unlike centers of mapping in Europe during this period, the Qing court showed little interest in sponsoring the production of maps of the whole globe. Yet the fact that it chose to map the claims of its empire in the ways that it did is significant. The practice of mapping is itself, of course, not new in this period, but several noteworthy developments occurred, namely the mapping of the entire Qing empire so that it could all be taken in by the viewer at once, and the adoption of the latest (international) technology both in undertaking the surveys and scientifically depicting the information obtained on the maps. While nonscaled maps with a substantial textual component continued to predominate, it is no coincidence that the representational conventions chosen for the atlas of the empire belonged to an emerging international map rhetoric—one that did not require schooling in a specific language, other than that of science, to understand. While place names still needed to be indicated in Chinese, or Manchu (or some other language),[1] the lands depicted on these maps, because drawn to scale, required no further textual explanation. Adopting the international conventions of precise measurement and representation qualified the new Qing maps to stake out territory internationally in a way that would not

---

1. The copy of the 1719 scroll map in the British Library (K.top.116.15, 15a, and 15b) attributed to Matteo Ripa has some place names and notes penned in Italian, in addition to the Chinese characters that label places within the provinces of China and the Manchu labeling of the territory beyond.

have been possible using earlier Chinese mapping practices; foreigners could grasp the significance of these maps at a glance without being schooled in either Chinese or Manchu language.

PARTNERS IN GEOGRAPHICAL LEARNING

Direct contact between European intellectuals and Chinese literati began at the turn of the seventeenth century when Jesuit missionaries first entered China. From the beginning they brought with them a marked faith in science as well as their religious convictions; no distinction was made between truth arrived at through science and the truth of religion. One of their goals was to learn more about world geography. Matteo Ricci, and later Martin Martini, arguably learned as much from the Chinese during this period as the Chinese learned from them about Europe; their maps of Asia were based almost entirely on Chinese sources.

Matteo Ricci's World Maps

As far as we know, the first European world map to be widely disseminated in China was one made by the Jesuit Father Matteo Ricci.[2] In 1583 Ricci was among the first Jesuits to enter China from Macao, where he had spent approximately one year studying Chinese language. He made his way to Canton, where he bestowed carefully selected gifts, including a self-chiming clock, in an effort to secure permission to reside

2. Ricci's Chinese world maps have generated much scholarly interest, including: John F. Baddeley, "Father Matteo Ricci's Chinese World Maps, 1584–1608," *Geographical Journal* 50 (1917): 254–70; Kenneth Ch'en, "Matteo Ricci's Contribution to, and Influence on, Geographical Knowledge in China," *Journal of the American Oriental Society* 59 (1939): 325–59, and 509 (for errata), and "A Possible Source for Ricci's Notices on Regions near China," *T'oung Pao* 34 (1938): 179–90; Marcel Destombes, "Une carte chinoise du XVIᵉ siècle découverte à la Bibliothèque Nationale," in *Marcel Destombes: Contributions sélectionnées à l'histoire de la cartographie et des instruments scientifiques*, ed. Günter Schilder, Peter van der Krogt, and Steven de Clercq (Utrecht: HES Publishers; Paris: A.G. Nizet, 1987), 469, and "Wang P'an, Liang Chou et Matteo Ricci: Essai sur la cartographie chinoise de 1593 à 1603," in *Appréciation par l'Europe de la tradition chinoise: À partir du XVIIᵉ siècle: Actes du IIIᵉ Colloque International de Sinologie* (Paris: Les Belles Lettres, 1983), 47–65; Pasquale M. D'Elia, *Il mappamondo cinese del P. Matteo Ricci S.I.* (Città del Vaticano: Biblioteca Apostolica Vaticana, 1938), and "Recent Discoveries and New Studies on the World Map in Chinese of Father Matteo Ricci, S.J.," *Monumenta Serica* 20 (1961): 82–164; Lionel Giles, "Translations from the Chinese World Map of Father Ricci," *Geographical Journal* 52.6 (1918): 367–85, and 53.1 (1919): 19–30; and Boleslaw Szczesniak, "Matteo Ricci's Maps of China," *Imago Mundi* 11 (1955): 127–36. On Ricci more generally, see Jonathan D. Spence, *The Memory Palace of Matteo Ricci* (New York: Penguin Books, 1984), and "Matteo Ricci and the Ascent to Peking," in *East Meets West: The Jesuits in China, 1582–1773*, ed. Charles E. Ronan and Bonnie B. C. Oh (Chicago: Loyola University Press, 1988).

in Guangdong province under the protection of the governor-general.[3] When the governor-general was reassigned to a new post five months later, Ricci had to renew his efforts to secure permission to maintain residence.

During his stay in Canton, Ricci devoted himself to learning classical Chinese, immersing himself in the Chinese classics, and fostering contacts with the literati. One area of common interest turned out to be geography. As early as 1584 a world map that Ricci had made and attached to the wall in his living quarters attracted the interest of his Chinese visitors. At the request of Wang Pan, an official who had admired the map, Ricci drew up a Chinese version. Later, the same man had it engraved at his own expense and presented copies to his friends, both in and beyond Guangdong province.[4] No copies of this version of Ricci's map are extant, but later editions, which included refinements and new information, followed. The most studied version of Ricci's world map is one he made together with his Chinese colleague Li Zhizao that was printed in Peking in 1602.[5] The last edition made during his lifetime appeared in 1608.

How much influence Ricci's map had has been debated, but we do know it received fairly widespread attention and circulation. The governor-general of Guizhou, for example, reproduced a copy in a book he wrote on Guizhou that was published in Guiyang in 1604.[6] By Ricci's own estimate, more than one thousand copies of the 1602 version were reprinted.[7]

3. Carlo M. Cipolla, *Clocks and Culture, 1300–1700* (New York and London: W. W. Norton and Company, 1978), 81. The choice of a device designed for measurement hardly seems coincidental.

4. Ch'en, "Matteo Ricci's Contribution," 343–44; Destombes, "Une carte chinoise du XVIᵉ siecle," 469. In 1594 Wang Pan published a map of China. In it, he says, one has "a relatively exact image of reality," and from it one can see that "the territory of the celestial court is larger than ever" (quoted from Destombes, "Une carte chinoise du XVIᵉ siecle," 468).

5. On the different editions, see D'Elia, "Recent Discoveries and New Studies." The 1602 edition is the only extant version. One copy is housed in the Vatican Library, one at Kyoto University, a third at the Library of Migayi Prefecture in Sendai, Japan, and a fourth in private hands. A facsimile edition can be found in D'Elia, *Il mappamondo cinese.*

6. Ricci had gotten to know Guo Zizhang (who also went by the names of Guo Qingluo and Bin Yisheng) during his early years in Guangdong when the latter was serving as prefect of Chaozhou in the eastern part of the province from 1582–1586. D'Elia, *Il mappamondo cinese,* 73–81, and "Recent Discoveries and New Studies," 96–107. See also Destombes, "Wang P'an, Liang Chou et Matteo Ricci," 59.

7. Berliner Festspiele, *Japan und Europa, 1543–1929* (Berlin: Argon, 1993), 228. Arthur Hummel, "Beginnings of World Geography in China," *Annual Report of the Librarian of Congress for the Fiscal Year Ended June 30, 1938* (Washington, D.C.: U.S. Government Printing Office, 1939), 224–26,

Not everyone was enamored of Ricci's map or of his science; some Ming scholars felt that his map purposely misrepresented the distance between China and Europe in order to minimize the appearance of any threat that European countries might represent. According to one critic:

> When these savage people [the Jesuit missionaries] say falsely that they come from a great distance of 90,000 *li*, it is because they want us to believe that they have no ulterior motives behind them, so as to prevent us from worrying about their aggressive purposes. In fact, these savages are full of subtle designs and resourceful schemes. Whenever they arrive at a country they would destroy her. Altogether they have subdued over thirty countries by direct conquest. It is difficult to examine the traces of their distant conquests, but we need just refer to their latest conquests of such regions as the Philippines . . . whose kings they killed and whose people they robbed. It required only a few of them to subdue an entire country. Are not these facts obvious proofs of their aggressive nature?[8]

While Kenneth Ch'en points to a confusion evident here between the Jesuit missionaries in China and Spain's activities abroad, I would suggest that the conflation of all Europeans into a more general category was not entirely misguided. Future events did bear out some of these fears.

In any case, Ricci's world map brought a new conception of the earth to the Chinese literati. Most important may have been the use of meridians to divide the globe into equal parts. Earlier Chinese maps, while sometimes representing territory beyond China, had not graphically conceptualized the different landmasses of the earth in proportion, or in strict relation, to one another. Seeing China as occupying only a relatively small part of the earth was a revelation. Other contributions of Ricci's map included the translation of geographic terms into Chinese, a record of recent discoveries made by Western geographers, and the

details five late Ming publications housed in the Library of Congress that contain copies of Ricci's world map. For distribution of Ricci's maps, see also Ch'en "Matteo Ricci's Contribution," 344–46. For subsequent publication of Ricci's maps in later sources, see Ch'en, idem, 346–47; Howard L. Goodman, "Paper Obelisks," in *Rome Reborn: The Vatican Library and Renaissance Culture*, ed. Anthony Grafton (Washington, D.C.: Library of Congress, in association with the Biblioteca Apostolica Vaticana, 1993), 261–62; Bernard H. K. Luk, "A Study of Giulio Aleni's Chih-fang Wai Chi," *Bulletin of the School of Oriental and African Studies, University of London* 40.1 (1977): 58–84; Arthur Hummel, "Astronomy and Geography in the Seventeenth Century," *Annual Report of the Librarian of Congress for the Fiscal Year Ended June 30, 1938*, 226–27. Some of these copies deviate in significant ways from Ricci's map.

8. Su Jiyu as quoted in Ch'en, "Matteo Ricci's Contribution," 349. For additional late Ming expressions of worries about the European threat, see Cordell Yee, "Traditional Chinese Cartography," 171–72.

image of a spherical earth.[9] Li Zhizao's Chinese preface to the map details many of these innovations for the literati audience. Ricci's map drew on both Chinese and European sources,[10] as well as his own surveys. As a result, his research allowed not only for better knowledge of geography in China, but also for a more accurate representation of Asia to be transmitted to Europe.

Like many maps of his day, both Chinese and European, Ricci's world map contained a substantial amount of text along with the graphic image of the world (see figure 8). The text he included can be divided into four categories. The first consists of a series of prefaces penned both by Ricci himself and by Chinese scholars, including Li Zhizao. The other texts deal with aspects of astronomy and cosmography; place names; and "notes on the products of various countries, the manners and customs of the inhabitants, etc."[11]

This last category of text reminds us that while Ricci's map was shaped by the best science of his day, it still belonged to an era when European maps regularly included information on the inhabitants of the countries depicted and when much of this information came from earlier literary sources and was not confirmed through direct observation. We learn, for example, that

> [i]n the Country of Disembodied Spirits just within the Arctic zone on the 90th meridian, "the natives go abroad by night and conceal themselves by day. They flay deer and clothe themselves in the skins. Their ears, eyes, and noses are like those of other men, only their mouths are on the top of their heads. They feed on deer and snakes."[12]

Like his information pertaining to the shape of the lands depicted, Ricci's textual sources are derived from both European and Chinese works.[13] Neither set of sources included images that were more fantastic or, by contrast, more empirical than the other. In Ricci's time, there was not yet a clear distinction within the field of geography between cartography and ethnography in either European countries or in Asia.

9. See Ch'en, "Matteo Ricci's Contribution," for a full analysis of Ricci's contributions to geographical knowledge in China. For the debate in China over the shape of the earth, see Pingyi Chu, "Trust, Instruments, and Cross-Cultural Scientific Exchanges: Chinese Debate over the Shape of the Earth, 1600–1800," *Science in Context* 12.3 (1999): 385–411.

10. Goodman, "Paper Obelisks," 263. Giles, "The Chinese World Map of Father Ricci," 52.6: 369; 53.1: 28.

11. Giles, "The Chinese World Map of Father Ricci," 52.6: 367, 369.

12. Ibid., 381.

13. Ibid., 381–82.

Furthermore, reports of "strange" peoples or customs were taken on the authority of textual sources. By the eighteenth century, by contrast, texts (and illustrations) describing peoples were, generally speaking, no longer found on European maps, but appeared in separate ethnographic genres such as descriptive geographies. By that time these depictions had also evolved to be based on direct observation rather than being based on hearsay or simply borrowed from older accounts. In China, while some maps still included ethnographic depiction, it was not appropriate to certain map genres.

In 1601 Ricci had gained permission to reside in Peking, where he remained until his death in 1610. In this context he continued to cultivate contacts with learned officials, and worked to facilitate scholarly exchange between China and Europe. He "requested that new missionaries should have had mathematical and astronomical training," [14] in order to prove useful to the Chinese court. But, he also felt that Europe should not be ignorant of China's rich scholarly heritage. To this end he translated the Chinese "Four Books," which contained the fundamentals of Chinese thought, for a European audience.

Giulio Aleni's Geographical Discourses

While Ricci remains the most well-known Jesuit to reach Ming China, he was not the only one. Although my purpose here is not to outline the early Jesuit presence in detail, it is appropriate to introduce other figures who were active in the exchange of geographical learning with Chinese literati. Besides Ricci, most prominent during the late Ming was Giulio Aleni. Later, Ferdinand Verbiest was also involved in cartographic pursuits.

Aleni was born in 1582, entered China c. 1613 after a stay of two to three years in Macao, and remained there through the Ming-Qing transition until his death in 1649. He had extensive contact with Chinese scholars and became an accomplished linguist, writing several works in classical Chinese. These included *Zhifang wai ji* (An Account of Countries Not Listed in the Records Office) and *Xifang da wen* (Questions and Answers about the West). His *Zhifang wai ji*, published in 1623, consisted of both a world map and an accompanying geographical treatise in which he described different regions of the world and the customs of their inhabitants. The title refers to the broad geographical

---

14. Bernard H. K. Luk. "Thus the Twain Did Meet? The Two Worlds of Giulio Aleni" (Ph.D. diss., Indiana University, 1977), 8.

scope of the work, indicating that it deals with countries outside the purview of the *zhifang si,* a Ming (and Qing) office in charge of maps.[15] The work has been called "the first world geography written in classical Chinese."[16] Although this may be an exaggeration, it was the first descriptive geography in Chinese to include a complete map of the world as projected onto a planisphere.[17]

The map it contained was based on Ricci's world map, and the texts compiled from several sources. The most important source seems to have been a manuscript copy of notes written earlier by two European Jesuits, Pantoja and de Ursis, at the behest of the emperor to augment Ricci's world map of 1601.[18] The content is very much in the spirit and style of other European geographical works dating from the same period.[19] However, in form and in the kinds of interests pursued, it also bears a resemblance to earlier Chinese treatises.[20] The content is, however, distinct; this was not a copy of a Chinese work. What I mean to suggest is that at this period the interests of European and Chinese geographers and the parameters within which they worked were not as different as one might expect. This is confirmed by the great interest that Aleni's geographical work sparked in China. The work was reprinted numerous times in later collectanea, including the prestigious *Siku quanshu* (commissioned in 1772). According to Bernard H. K. Luk, it was "the best remembered of all [Aleni's] books among the Chinese."[21]

Xifang da wen, Aleni's other well-known geographical work, was published in 1637.[22] Taking the form of questions and answers, it

15. Also translated as "Bureau of Operations," the *zhifang si* was one of four bureaus in the Ministry of War. See Charles O. Hucker, *A Dictionary of Official Titles in Imperial China* (Stanford: Stanford University Press, 1985), 981.

16. Luk, "Giulio Aleni's Chih-fang Wai Chi," 61. Luk's article also includes translations of portions of the text.

17. It was not the first showing regions beyond China. See Marcel Destombes, "Une carte chinoise du XVIe siècle," 474; and Walter Fuchs, *The "Mongol Atlas" of China,* Monumenta Serica, monograph no. 8 (Peiping: Fu Jen University Press, 1946).

18. George H. Dunne, *Generation of Giants: The Story of the Jesuits in China in the Last Decades of the Ming Dynasty* (Notre Dame, Ind.: University of Notre Dame Press, 1962), 115.

19. Specifically, Sebastian Münster's *Cosmographiae Universalis Libri VI* (1550, 1572), Peter Heylyn's *A Little Description of the World* (sixth ed., 1633), George Abbot's *A Briefe Description of the Whole World* (1605), and Ortelius, *Il Teatro del Mondo* (1598). Luk, "Giulio Aleni's Chih-fang Wai Chi," 78.

20. Ma Huan's *Ying yai sheng lan,* for example. For more on Chinese geographical treatises of the Ming period, see Needham, 558–59.

21. Luk "Giulio Aleni's Chih-fang Wai Chi," 63, 80.

22. See John Mish, "Creating an Image of Europe for China: Aleni's *Hsi-fang Ta-wen,*" *Monumenta Serica* 23 (1964): 1–87, for a complete translation.

provides a general introduction to European society and values to the reader. The subheadings include: (in book 1) geographical data, traveling routes, ships, dangers of the sea, pirates, sea marvels, landing (i.e., disembarking), products, manufactures, kings, Western learning, offices and titles, dress and ornaments, customs and usages, the five social relations, legal system, entertaining guests, commerce, food and drink, medicine, character, welfare institutions, palaces and houses, walls, moats, and military works, marriage (and second marriages), remaining chaste, burial rites, mourning clothes, funerals, and sacrificing to ancestors; and (in book 2) maps, astronomical calculations, eclipses, constellations, years and months, beginning of the year, year designations, and Western scholars. An appendix addresses geomancy.[23] The topics chosen reveal a sensitivity on the part of Aleni to the kinds of questions that concerned the Chinese literati. Ancestor worship, the five social relationships, and geomancy are not categories one would normally expect to find in descriptions of Western countries; their inclusion reflects the concerns of the Chinese readership.

The introduction to *Xifang da wen* was written by Li Zhizao, the same scholar who wrote a preface for Ricci's map. His words illuminate for us the interest a Chinese reader brought to the work, and the criteria by which the work might have been judged. He states:

> All sorts of unexpected information [in this book] are delightful and surprising. Although we had never before heard the like of it, everything is based on [the missionaries'] personal experience or . . . documented in their countries.[24]

In his view the book was not only engaging because of the curiosity of its contents, but valuable due to the method by which the observations were compiled. Personal experience and careful documentation were the criteria of the Chinese literati audience, most of whom would have identified with the *kaozheng* movement. To this extent its participants shared in the same kind of epistemological outlook as their European counterparts working at the court who were paying ever increasing attention to what might be referred to as scientific method.[25]

---

23. I have retained the subheadings as translated by Mish.

24. Quoted in Luk, "Giulio Aleni's Chih-fang Wai Chi," 75, note 24.

25. An emphasis on measurement and even "realism," while fundamentally related to our definition of the early modern period, was not unique to early modern Europe. Although the topic is only beginning to be explored, interesting work is being done in this area in late Ming and Qing China. At the 1997 meeting of the Association of Asian Studies, March 12–16, 1997, numerous papers were devoted to arguing for evidence of an epistemology rooted in observation and measurement during this period. See, for example, Kenneth Ganza, "To Hear with the Ears Is Not as

Aleni's career path never took him to the capital, and he did not serve in the direct employ of the Chinese bureaucracy. Rather he settled in Fujian province, where he became something of a well-known figure. In some ways he may have had more opportunity to cultivate relationships with Chinese colleagues than his counterparts in Peking. We know that he had extensive contacts with Chinese literati, many of whom were officials. In 1613 he traveled with Xu Guangqi, a wealthy Christian convert, to Shanghai.[26] In 1620, when he was writing *Zhifang wai ji*, he stayed with a Chinese friend, and retired official, by the name of Ye Xianggao.[27] He also knew and shared common interests with Jiang Dejing, who helped him to edit *Xifang da wen*. Jiang later became president of the Board of Rites, and served for a time as a Grand Secretary. Jiang himself authored a book "on the military and economic geography of the Chinese border regions."[28] Aleni was no stranger to Chinese intellectual circles; many members of the late Ming intelligentsia shared common interests with him.

Although Aleni did not live or work in the capital, ever since Ricci had made inroads there other Jesuits did perform services for the imperial court. Johann Adam Schall von Bell (1591–1666) and Giacomo Rho (1593–1638), for example, translated Western books on astronomy and calendrical systems by imperial order.[29] Schall was eventually made head of the Bureau of Astronomy, a post that would subsequently be held by Ferdinand Verbiest (1623–1688).[30]

Verbiest continued the practice of making world maps. His patrons were, however, the Manchus, whose Qing dynasty had superseded the Ming in 1644. In 1674 he made a world map for the Kangxi emperor.[31]

Good as to See with the Eyes: Travel Painting as an Expression of Empiricism in the Late Ming and Early Qing"; Leo K. Shin, "The Culture of Travel Writings in Late Ming China"; John Herman, "The Cant of Conquest: Creating 'Barbarians' and 'Chinese' in the Southwest"; and James Millward, "Mapping Land and History: Qing Depictions of Xinjiang/the Western Regions." A revised version of Millward's paper has since been published under the title "'Coming Onto the Map': 'Western Regions' Geography and Cartographic Nomenclature in the Making of Chinese Empire in Xinjiang," *Late Imperial China* 20.2 (December 1999): 61–98.

26. Luk, "Giulio Aleni's Chih-fang Wai Chi," 62.

27. Ibid., 62.

28. Mish, "Creating an Image of Europe for China," 2.

29. Ibid., 71.

30. For more on Schall and Verbiest, see Jonathan D. Spence, *To Change China: Western Advisers in China, 1620–1960* (Boston and Toronto: Little, Brown and Company, 1969), chapter 1.

31. R. V. Tooley, *Maps and Map-Makers*, 4th ed. (New York: Bonanza Books, 1970), 106. Theodore N. Foss, "A Western Interpretation of China: Jesuit Cartography," in Ronan and Oh, *East Meets West: The Jesuits in China, 1582–1773*, 214. The map was entitled *Kunyu quantu*, or "Complete Map of the Terrestrial Globe."

It was not simply a copy of Ricci's map, but was instead based primarily on the more recent 1661 map of Nicolaus à Wassenaer.[32] He also authored two textual works in Chinese dealing with the world beyond China. *Kunyu tushuo* (Comments on the Map of the World) appeared in 1674. As its title indicates, it was a kind of companion piece to the world map, much as Aleni's *Zhifang wai ji* had been a companion piece to Ricci's map. In fact, although more detailed, it was based on Aleni's work. The other book, *Xifang yao ji* (Main Points about the West), took Aleni's *Xifang da wen* as a model.[33] As with several of Aleni's geographical works, the *Kunyu tushuo* was reprinted in later Chinese collectanea.[34]

China in European Maps before 1700

Prior to the sixteenth century, maps of China were no more likely to be found in Europe than maps of Europe in China. Portuguese exploration and Jesuit missions gradually led to an interest in and need for maps of Asia. In 1584, very nearly the same time Ricci presented his first world map to a small circle of Ming literati, Ortelius included a map of China in his *Theatrum Orbis Terrarum*[35] (figure 12). Still far from an accurate representation of China's actual landmass, it was a beginning. Although Ricci soon sent more accurate cartographic representations back to Europe,[36] the one printed by Ortelius remained the standard for more than sixty years.[37]

Not until the middle of the seventeenth century did the shape of China on European maps undergo significant improvement in terms of representational accuracy. In 1655 Martino Martini's *Novus Atlas Sinensis,* based on research he carried out as a Jesuit missionary in China, revised the shape of European maps of China. His work was based largely on Chinese maps of China, not on surveys.[38] He relied on the

32. Foss, "A Western Interpretation of China," 214.
33. Luk, "Giulio Aleni's Chih-fang Wai Chi," 76.
34. Ibid., 40, 80.
35. Additamentum III. Copies of both Ortelius's *Civitates Orbis Terrarum* and his *Theatrum Orbis Terrarum* reached China in the late Ming. See Harrie Vanderstappen, S.V.D., "Chinese Art and the Jesuits in Peking" in *The Jesuits in China, 1582–1773,* ed. Charles E. Ronan, S.J., and Bonnie B. C. Oh (Chicago: Loyola University Press, 1988), 108. For more on European maps of China, see Zögner, *China cartographica.*
36. Szczesniak, "Matteo Ricci's Maps of China," 131.
37. Tooley, *Maps and Map-Makers,* 106; Foss, "A Western Interpretation of China," 211.
38. Although Ricci had taken certain latitudinal sitings for improved accuracy in his map, this kind of activity was limited (in both quantity and accuracy) until the early eighteenth century.

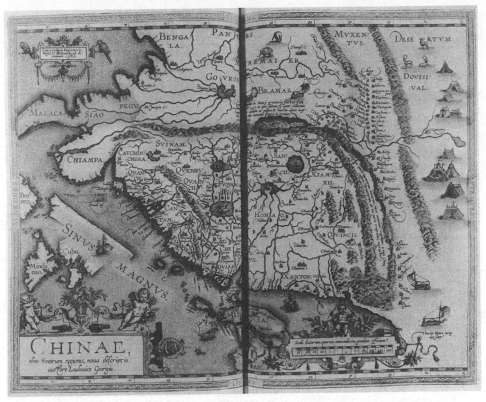

Figure 12 Ludovico Georgio's Map of China, from Martino Ortelius's *Theatrum Orbis Terrarum*, Antwerp, 1584. Courtesy The Newberry Library, Chicago.

Guang yu tu most directly, but also on local gazetteers and on leads his Chinese colleagues provided. His work was, as the title suggests, not a single map, but a whole atlas of China. It contains one small-scale map of all of China (figure 13), and also a map for each of the fifteen provinces of the Ming dynasty, as well as a map of Japan. Not limited to maps, the work also includes textual descriptions of different parts of the country: "He gave statistical, physical, economic, and political geographic information on each of the provinces."[39] The maps themselves, like Ortelius's earlier map, were also illustrated, imparting information on costume and products of the lands depicted. Unlike Ortelius's map these illustrations were, however, largely confined to

39. Foss, "A Western Interpretation of China," 216.

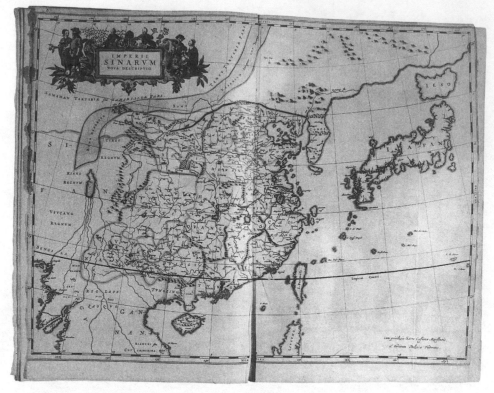

Figure 13 Map of China, from Martino Martini's *Novus Atlas Sinensis*, 1643. Courtesy The Newberry Library, Chicago.

the borders and to the cartouches rather than placed on the maps themselves. Martini's image of China superseded earlier representations, and remained the standard in Europe until d'Anville published his *Atlas de la Chine* (which was based on the Kangxi Jesuit surveys) in 1737[40] (see figure 14). So admired was Martini's work that, according to one sinologist, it contributed to Louis XIV's decision to send mathematicians to China.[41]

During the seventeenth century China and Europe shared a mutual interest in cartography, and in sharing cartographic information. Neither had a monopoly on accuracy.

40. Tooley, *Maps and Map-Makers*, 107.
41. Destombes, "Une carte chinoise du XVIᵉ siecle," 478.

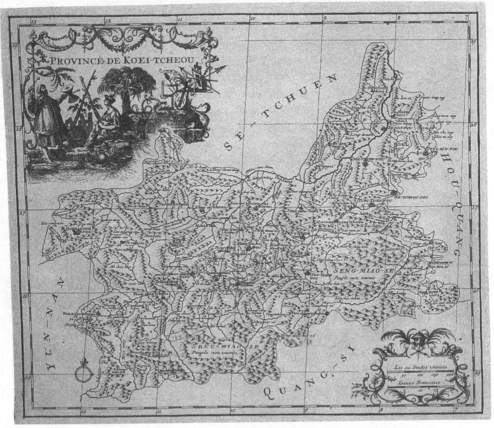

Figure 14 Map of Guizhou Province, from d'Anville Atlas, 1737. Courtesy The Newberry Library, Chicago.

ADOPTION OF EARLY MODERN MAPPING TECHNIQUES IN FRANCE, QING CHINA, AND MUSCOVY

The Kangxi atlas project followed, and in some sense built on, indigenous cartographic projects. At the same time it also fit into Europe's search for more precise geographical information worldwide. The majority of the European Jesuits who were involved in the Kangxi surveys came from France, a country experiencing the dual impetus for the creation of national maps and the larger, but related, project of mapping the globe. The latter project, although it required international cooperation, was not without national rivalries among the participants.

In the Qing context it is necessary to explore the combination of factors that the rulers had to juggle. Their own Manchu identity, maintaining legitimacy in the eyes of the Chinese, and technological innovation necessary for continuing military success had to be weighed against each other to find the best balance for the continued health and viability of the dynasty. We must also carefully consider the extent of the Kangxi emperor's own geographical knowledge and the related matter of how much importance he placed on both Qing and world geography. This question can be explored from several angles, including the range over which his personal knowledge of and interest in early modern technologies of representation extended, how proactive he was in recruiting and employing skilled (foreign) technicians, and to what extent he limited their access to what he considered to be the most sensitive information that would be included in the atlas. The international nature of the Qing surveying project also involved both transcontinental cooperation and competition. The contradictory needs to guard against the indiscriminate dissemination of highly sensitive geographic information and to publicize territorial claims abroad were in constant tension. Finally, we must also consider what purposes the atlas was to serve. Was it for wide circulation or to be carefully protected from unauthorized eyes? These questions have implications for the dissemination, or lack thereof, of the new mapping technologies and of the various versions of the Kangxi atlases themselves.

The French Context

A recent collection of essays on mapping in Europe, entitled *Monarchs, Ministers, and Maps,* explores the emergence of cartography in the early modern period.[42] The volume makes it clear that, although mapping had an important role in the development of Europe in this period, and that over time to-scale mapping based on surveys eventually became crucial throughout Europe, there is no unified "European" or "Western" cartography. Different countries took the lead in mapping technology at different times, each contributing distinctive characteristics to its own national cartography through the specific uses made of the maps and the features and technologies incorporated in their production. Generally speaking, the Italian city-states, where Ptolemy's *Geographia* was first reprinted, took the initial lead.[43] They were followed in this role by the

---

42. Buisseret, *Monarchs, Ministers, and Maps.*
43. Tooley, *Maps and Map-Makers,* 19.

Netherlands, France, and Britain.[44] As Tooley has stated, "Map-making of the past on a large scale has always followed sea power and wealth, and the loss of either to a consequent decline."[45] In China too, the height of cartographic endeavor in the Qing dynasty paralleled the height of imperial (if not sea) power. China's cartographic endeavors are best viewed not in contradistinction to "a" European mapping tradition, but rather examined along with contemporaneous advances in the different European centers.

This discussion of mapping in Europe concentrates on France for two reasons. First, most of the foreign surveyors to participate in making the Kangxi atlas were French and had received their training in France. Secondly, during the late seventeenth and early eighteenth centuries, France was Europe's leader in terms of cartographic activity and therefore makes the best point of comparison for China's cartographic activity during the same period. Both French and Ming dynasty rulers had had their realms, or extensive portions thereof, mapped in the late sixteenth and seventeenth centuries, but it was not until the late seventeenth and early eighteenth centuries that precise surveys were initiated in both places.

Cartography in France was first practiced extensively under Catherine de Médicis, the widow of Henry II (r. 1547–1559). Circa 1560 she commissioned Nicolas de Nicolay (1517–1583), to "draw up a general account of France, accompanied by maps of each province."[46] It is interesting to note that as with the Ming gazetteer of China *(Da Ming yi tongzhi)*, textual descriptions formed the basis of the work, which was only embellished by maps. Catherine also recognized the utility of maps for overseas ventures. Later under Henry IV (r. 1589–1610), who shared her interests in mapping, more maps were produced, both of France and of regions abroad where France had interests.[47] The most sensitive of the military maps produced were not published, but remained in manuscript for a limited number of eyes. Under Louis XIII Cardinal

---

44. This gross outline is not meant to preclude recognition of John Speed's maps of Britain in the sixteenth century or Spain's achievements, but simply to highlight how the centers of activity shifted over time.

45. Tooley, *Maps and Map-Makers*, 47.

46. David Buisseret, "Monarchs, Ministers, and Maps in France before the Accession of Louis XIV," in Buisseret, *Monarchs, Ministers, and Maps,* 106.

47. For more detail on mapping under Henry IV, see Buisseret's chapter in *Monarchs, Ministers, and Maps,* and François de Dainville's, *La géographie des humanistes* (Paris: Beauchesne et Ses Fils, 1940). Dainville is particularly explicit about Henry IV's designs overseas. Most famous of the French maps of overseas made during his reign are Samuel de Champlain's maps of New France.

Richelieu continued to sponsor mapping projects both for military and other uses, but it was not until the reign of Louis XIV (1643–1715), which was largely contemporaneous with the Kangxi emperor's reign (1662–1722), that topographical surveys of all of France based on astronomical observations would be initiated under the guidance of Jean Dominique Cassini, a foreigner who was later naturalized. During the late seventeenth and early eighteenth centuries France took the lead in early modern cartographic practice.

Jean-Baptiste Colbert (1619–1683), minister for home affairs under Louis XIV, was the leading figure in establishing France as a center for the study of science, including cartography. In fortifying France's position as a leader in affairs both military and scientific, Colbert reached beyond the confines of the realm to employ talent from all over Europe. He sent personal invitations to the best minds in Germany, the Netherlands, Italy, and England, offering substantial pensions as incentive to attract them into the service of Louis XIV.[48] To solidify France's leadership role he founded the Royal Academy of Sciences in 1666. One of its primary functions was to make astronomical observations crucial to improved navigation and mapping. Acutely aware of the conditions in observatories in other countries, Colbert was determined that the Paris Observatory "should surpass in beauty and utility any that had been built to date, even those in Denmark, England and China," and "reflect the magnificence of a king who did things on a grand scale."[49] Leadership in the sciences was a matter of pride as well as pragmatism.

From his accession to power, Colbert recognized the necessity for good maps of the realm. In 1663, even before the possibility of a complete topographical survey was seriously contemplated, he

> instructed officials in the provinces to survey the condition of France. He mentioned maps at the beginning of his instructions, a very lengthy memorandum that touched upon every kind of administrative unit and government function. He asked that the crown's agents send the capital accurate and detailed maps of each province and généralité and have new maps made by capable, intelligent, and skillful persons of areas not covered satisfactorily by existing maps. He further specified that written reports be submitted to describe areas that no map, new or old, covered and that maps and reports alike illustrate the divisions between military, judicial, fiscal, and ecclesiastical administrative units.[50]

48. Lloyd A. Brown, *The Story of Maps* (New York: Dover Publications, 1979), 214.
49. Ibid., 214. See also C. Wolf, *Histoire de l'Observatoire de Paris de sa fondation à 1793* (Paris: Gauthier-Villars, 1902), 4.
50. Konvitz, *Cartography in France, 1660–1848*, 1–2.

The aim was through a combination of direct observation and existing records to collect as much information as possible so that the central government might be able to know better the conditions of the realm it was to govern.

Only a few years later the ability to measure longitude from land would transform cartographic practice. In 1668 Cassini, then still in the employ of Pope Clement IX, published his *Ephemerides*—a series of tables based on carefully timed accounts of the eclipses of the moons of Jupiter. With that work, as revised and enlarged in 1676, astronomers could determine the latitude and longitude of the point from which they made observations.[51] From this time mapping became a global endeavor; every point that could be measured was of both theoretical and practical necessity charted in relation to other points on the globe. Practically speaking national maps would no longer exist as independent entities, because even if they showed no territory beyond the confines of a given state, the lines of latitude and longitude in relation to which the land was placed spoke quietly of what lay beyond. Maps of all of Europe would need to be redrawn based on this new standard of accuracy.

Immediately recognizing the potential of Cassini's contribution, Colbert made every effort to attract him to France to work in the service of the Academy of Sciences under Louis XIV. Cassini arrived in France in 1669, where he would spend the rest of his life. Now the twofold effort to survey France and to map the globe with accuracy could begin in earnest.

Even before Cassini's arrival in France measurements based on his tables had gotten underway. In 1668 scientists from the Academy of Sciences interested in developing and testing ways of obtaining more accurate cartographic measurements surveyed a limited amount of territory near Paris. A decade later, in 1679, members of the same group obtained approval from Louis XIV to extend the survey to "make a map of France with the greatest possible accuracy."[52] Astronomical observations made in conjunction with Cassini's tables would be used to determine exact measurements for locations in France. Key to the methodology employed was the implicit understanding that the "size and form of France could be determined only once the size and shape of the earth

51. The technology to determine longitude from sea was yet to be developed. An Englishman, John Harrison, would later invent the marine chronometer solving this thorny problem and making much more accurate navigation possible.

52. Konvitz, *Cartography in France, 1660–1848*, 5.

itself were known."[53] Mapping France in this way constituted a realization that it occupied only one part of a finite globe, and that other points on the globe and the boundaries between countries could be demarcated in the same way. France constituted one piece of the global puzzle.

Scientists at the Academy therefore began to chart the globe. The third floor of one whole tower of the new state-of-the-art observatory was devoted exclusively to this endeavor. Starting with a blank slate—or rather planisphere—points were gradually filled in as observations could be carried out internationally.

> What interested the Académie was the precise location, according to latitude and longitude, of the important places on the earth's surface, places that could be utilized in the future for bases of surveying operations. For this reason it was much more important to have the names of a few places strategically located and widely distributed, according to longitude, than it was to include a great many places that were scientifically unimportant. For the same reason, most of the cities and towns that boasted an astronomical observatory, regardless of how small, were spotted on the map.[54]

The achievements of the observatory were widely renowned; the king himself paid a visit to the grounds, where he witnessed demonstrations of the new techniques[55] (see figure 15).

The Academy's need for observations from around the globe brought together the interest of the crown and the goals of those who would establish a French mission presence in China. Although the Jesuit order had existed since the middle of the sixteenth century, and Jesuit missionaries were present in China from the later part of the same century, they were sponsored through Rome. Under Louis XIV, with the Academy of Sciences' need for direct channels of information from abroad, Louis's own desire for international recognition of the achievements of his court, and recognition of the opportunity this combination of factors provided to finance further missionary endeavors, the Missions Étrangères was formed in 1663. The king provided an annuity of 15,000 pounds.[56]

Long leading figures in education, science, and particularly geography, the French Jesuits were in an excellent position to make connec-

---

53. Ibid., 4.
54. Brown, *The Story of Maps,* 219.
55. Ibid., 220.
56. *La grande encyclopédie: Inventaire raisonné des sciences, des lettres et des arts,* vol. 23 (Paris: Société Anonyme de la Grande Encyclopédie, 1902), 1121.

Figure 15 Louis XIV's Visit to the Observatory. *Les animaux et les arts.* Courtesy The Newberry Library, Chicago.

tions between foreign centers and Paris. Although French Jesuit presence abroad was in no way limited to China, it is their inroads at the Qing court that interest us here.

The Qing Context

During the seventeenth century, before the court employed European technicians to make maps, we see evidence of a recognized need for more detailed maps. Resulting mapping efforts paralleled, but occurred independently of, increased mapping activity in both France and Russia during roughly the same period. The need for better maps was pursued in all three places before the new technology for measuring longitude from land, which facilitated accurate to-scale mapping, became available and was adopted in Europe and, for certain purposes, by the Russian and Qing courts.

From their accession to the emperorship of China, the Manchu rulers recognized a need for better records of land use and taxation. In 1646, only two years after the conquest, Dorgon, the regent, ordered a cadastral survey of the empire.[57] Administration could only be as effective as information was accurate. The empire-wide cadastral surveys were followed up in subsequent decades with imperially commissioned studies on a number of individual provinces. For example, in 1684, after various officials expressed dissatisfaction with the inadequacies of existing geographies of Guangdong, the Kangxi emperor commissioned a new geographical handbook of the province. As a result, "a systematic inspection was made in each administrative division of the province to gain more accurate information on the names and locations of mountains and streams, the natural and artificial boundaries, the historic and scenic places and the distances between them." The findings were submitted on ninety-seven new maps.[58] In purpose and design (although not physical appearance), the survey of Guangdong province seems not unlike the one commissioned by Colbert for all of France in 1663. Local gazetteers of Guizhou, which included both geographical and other detailed information on that province (discussed extensively in chapter 5), were also commissioned under the Kangxi emperor.

In China, as in France, it was not long before new surveying techniques made much more accurate depictions of the geographic features of the empire feasible. What is significant in this context is not so much

57. Yee, "Traditional Chinese Cartography," 177–78.
58. Arthur Hummel, "Atlases of Kwangtung Province," *Annual Report of the Librarian of Congress for the Fiscal Year Ended June 30, 1938,* 229–30.

the availability of the techniques developed in Europe as the imperial recognition of their utility and the wide sponsorship cartographic projects received. Although Frenchmen claimed credit for planting and nurturing within the Kangxi emperor the desire for more accurate maps,[59] the fact remains that Qing recognition of the utility of cartography predates their presence at the court. Ultimately the responsibility, or the credit, for the decision to map the empire using the most accurate and up-to-date techniques, and whom to employ in the endeavor, lay with the emperor himself, not those who actually carried out the project.

The decision to survey the empire and to employ European Jesuits to carry out much of the legwork was not made overnight, but based on a number of trials near the capital (not unlike the surveying of Paris in 1668, which predated the commissioning of the French national survey map under Colbert). The first project, to survey and map the environs of Peking, was commissioned and carried out in 1707. Pleased with the results, the emperor requested a second survey in 1708, this time of portions of the Great Wall. Finally, he ordered a survey of the whole of Beizhili, the province containing the capital. It was successfully completed to the emperor's satisfaction in 1710, and the Jesuit surveys of the rest of the empire soon followed.[60] Even Yee, who ultimately argues against significant Western influence on Chinese mapping traditions during the Qing, recognizes that "the Kangxi emperor turned to the Jesuits as an alternative to Chinese mapmakers," and that he "saw the initial stages of the survey as a kind of contest between cartographic traditions."[61] That the Kangxi emperor saw good reason to employ foreign technicians in this endeavor is not in question. As seen above in chapter 1 he was thoroughly aware of European expansion around his empire, including the most pressing threat from Russia to the north.

The Romanov Context

At roughly the same time as both France and the Qing, Russia also became heavily engaged in early modern mapping activity. Peter the Great (r. 1682–1725), like the Kangxi emperor (r. 1662–1722), was actively expanding his empire, and simultaneously recruiting talent to map it according to the latest techniques. A brief survey of mapping activity

59. Gaubil, *Correspondance de Pékin*, 214 (letter to P. Souciet, 1728); Foss, "Western Interpretations of China," 222–23.

60. For a contemporary, although brief, account of the surveys, see Matteo Ripa, *Memoirs of Father Ripa*, trans. Fortunato Prandi (London: John Murray, 1844).

61. Yee, "Traditional Chinese Cartography," 184–85.

in Russia will help to demonstrate the competition to acquire accurate maps, as well as the international impetus for accurate maps that characterized the beginning of the eighteenth century. Furthermore, it may help to explain the Kangxi emperor's decision to limit direct Jesuit access to the most sensitive frontier regions of the Qing empire, as discussed further below.

During the sixteenth century, Ivan the Terrible, first czar of Russia, and other Russian rulers attempted to have maps made of their domains, but experienced only limited success.[62] A century later, in the 1660s, clear direction from the throne to collect geographical information on various parts of the empire came when Czar Alexei Mikhailovich commissioned a map of Siberia now known as the Godunoff Map. The map was produced not by survey, but by gathering as much data together as possible, from

> trappers, traders, and others, Russians and natives, speaking many tongues; where they had been, in what circumstances, with names of places and peoples, and of mountains and rivers, and, above all, the distances traversed; or, what was much more to the point, the number of hours or days it took to go from place to place in given conditions—towing or poling up-stream, rowing or sailing down; riding in summer, skiing or driving dog- or reindeer-sledges in winter.[63]

One cannot help but notice that the 1660s were an extremely active time cartographically in France, China, and Russia, even though the technology which would allow longitude to be calculated accurately from land was yet to become available.

In 1698 Peter the Great, desiring better maps of Siberia, commissioned Semion Remezov to carry out further research and to submit new maps. As in the earlier Godunoff map the information was based on existing maps and interviews rather than surveys. This work, an atlas consisting of twenty-four pages, was completed in 1701.[64]

---

62. Cracraft, *The Petrine Revolution in Imagery*, 271.

63. John F. Baddeley, *Russia, Mongolia, and China*, vol. 1 (New York: Burt Franklin, 1919), cxxv. The original is no longer extant. We know of it from two Swedish copies made in late 1668 or early 1669. One was sketched by Johansson Prutz, who was permitted to see it by Prince Ivan Alexeivich Vorotinski on the express condition that he not draw it. The other was made by Fritz Cronman. (Baddeley, cxxvi.)

64. Ibid., cliii. Remezov had also been commissioned to make a wall map of Siberia for Peter the Great in 1696. He completed it one year later (ibid., cxxxvii and clviii). For more on cartography in Russia during the early modern period, see Bernard V. Gutsell, ed., *Essays on the History of Russian Cartography, 16th to 19th Centuries*, trans. James R. Gibson (Toronto: University of Toronto Press, 1975), especially S. Ye Fel, "The Role of Petrine Surveyors in the Development of Russian

From these instances of late seventeenth-century French, Qing, and Russian mapping activities we can only conclude that technology alone did not create the need for accurate maps, nor did the need spread internationally from a European center only after improved technology became available. Rather, in France, Qing China, and Muscovy—all powerful and expansionist realms—need for better maps would appear to have been felt independently, but simultaneously. For this reason once superior technology for accurate mapping became available they all chose to adopt and foster it.

The earliest evidence we have of Peter the Great's direct contact with early modern mapping techniques dates from his 1697–98 embassy to Europe, when he commissioned a Russian map to be printed in Amsterdam. Based on information he provided, it showed the territory of "southern Russia, the Crimea, and the Black Sea, the location of Peter's current strategic interests and site of his Crimean campaigns of 1695–96 and resultant conquest of Azov." Both Russian- and Latin language editions appeared in 1699.[65] From this time cartographic activity continued in Russia at a lively pace, much of it carried out by non-Russian Europeans.

Perhaps the most famous of these was Joseph Nicolas Delisle, a French astronomer who spent over twenty years in Russia (1725–1747) as professor of astronomy at the St. Petersburg Academy of Sciences. Joseph Nicolas was one of three brothers all involved in scholarly pursuits. Peter the Great had first met his elder brother Guillaume, a member of the French Academy of Sciences and a mapmaker, in 1717.[66] Impressed with Guillaume's 1706 "Carte de Moscovie," Peter wanted his assistance in mapping the full extent of the Russian empire, the eastern borders of which were still undetermined.[67] This project would take many years, and was directed in large part by Joseph Nicolas after his brother's death. Actual surveys of the empire were not formally initiated until 1727 on the orders of Peter's successor Catherine the Great. The resultant work was published in 1745 under the title *Atlas Russicus*. The

Cartography During the 18th Century," and V. G. Churkin, "Atlas Cartography in Prerevolutionary Russia." See also Alexei V. Postnikov, "Outline of the History of Russian Cartography," in *Regions: A Prism to View the Slavic-Eurasian World: Towards a Discipline of "Regionology,"* ed. Kimitaka Matsuzato (Sapporo, Japan: Slavic Research Center, Hokkaido University, 2000), 1–49.

65. Cracraft, *The Petrine Revolution in Imagery*, 274.

66. In 1718 Guillaume Delisle was given the title "Premier Géographe du Roi" (Tooley, *Maps and Map-Makers*, 43).

67. Cracraft, *The Petrine Revolution in Imagery*, 276.

atlas appeared not only in Russian, but also in Latin, French, and German editions.[68]

CARTOGRAPHY AS A CONSTITUTIVE COMPONENT OF EARLY MODERN EMPIRES

We have seen that mapping projects took place independently in France, Qing China, and Russia during the 1660s. In terms of the individuals involved in carrying out the projects and in the derivation of the methods used at that time, there were no apparent connections between these projects. What the rulers of these countries did have in common was an increasing awareness of their own kingdom's position as one country located on a finite globe. All were striving to consolidate their rule within their domains, and simultaneously working towards establishing boundaries in an early modern world context where sovereignty was gradually becoming increasingly tied to territorial integrity.

Once the technology to carry out geodetic surveys based on accurate astronomical observations of both latitude and longitude was available, none of these countries lost time in making use of it. Nor did they hesitate to draw on expertise from outside their own borders. France was in many ways on the cutting edge; Colbert used every means to attract Cassini, who had made the breakthrough for measuring longitude from land, to the service of his court. Soon in France both domestic surveys and the ambitious plan to chart the globe were underway. Qing China, however, did not lag behind; in spite of its almost incomparably greater size, the Kangxi surveys of the empire were completed by 1717, twenty-seven years before the national survey of France in 1744. The Qianlong revisions to the survey were likewise completed earlier (1755) than the second edition of France's survey (1788).[69] As we have seen, the completion of the atlas of the Russian empire in 1745 closely followed that of the national survey of France.

In the case of Russia, a convincing argument has been made that

68. Ibid., 278.

69. Konvitz, *Cartography in France, 1660–1848,* 16. One might argue that the Kangxi surveys were not entirely complete as portions of the atlas (namely Tibet and Korea) were not based on direct surveys. However, as Josef Konvitz reminds us, it is also the case that "[d]evelopments in French cartography in the eighteenth century produced an image of the country which was more integrated and centralized than was the reality at the time." See Josef Konvitz, "The Nation-State, Paris and Cartography in Eighteenth- and Nineteenth-Century France," *Journal of Historical Geography* 16.1 (1990): 3. Neither survey was a perfectly accurate representation of reality; both nonetheless put forth important claims for the territorial integrity and sovereignty of their states.

Peter the Great, by using European advisers and adopting the latest early modern cartographic technologies, put Russia on the map of Europe.[70] Stated differently, Russia (although not other parts of what is now the former Soviet Union) conceptually became part of Europe during the eighteenth century, a transition symbolized by the concurrent name change from Muscovy to Russia. Peter's achievements have endured; Russia remains on the map of Europe today.

In the case of Qing China, early modern mapping technology was as effectively used in the process of enlarging and consolidating the empire. Because of its geographical remove from Europe, however, it could not be put on the map of Europe in the same way. Yet, China was at this time put on maps made and circulated widely in Europe. The shape of modern China, in the form of national borders recognized in the twentieth century, was largely determined through Qing territorial claims communicated via early modern cartography and its international dissemination in the eighteenth century. In this respect Qing use of early modern mapping technology was every bit as effective as Russian adoption of the same. We must not let a false dichotomy between East and West, culture and history, obscure this important fact.

Ironically, maps made under the Manchu Qing emperors took on a power of their own in laying the geographical foundation for the modern Chinese nation-state that succeeded it. Three versions of the original Kangxi Jesuit maps now housed in the British Library convey different nuanced understandings of what constituted China, on the one hand, and the Qing empire, on the other; the two entities were not coterminous. The largest edition, in three scrolls, clearly distinguishes China proper *(neidi)* from the rest of the Qing empire in two ways.[71] First, the borders of China's provinces are represented by dotted lines. This demarcation sets them off not only from each other, but from foreign countries to the south and southwest and from Qing territory to the north and west. Secondly, and more graphically indicative of China's position as only one part of the Manchu empire, place names within China's provinces are written in Chinese characters whereas territory beyond, including Liaodong, is labeled exclusively in Manchu. Thus we see China as one distinct part of the larger Manchu empire. Its status as a colonized territory that formed only one part of the empire stands out vividly on this map.

70. Cracraft, *The Petrine Revolution in Imagery.*
71. British Library (K.top116.15, 15a, and 15b).

Two other versions of the same surveys, which would have had somewhat broader although still limited circulation, were entirely in Chinese with no Manchu script.[72] This may have been done to facilitate their use among Han officials, or so as not to offend Han sensibilities by an implication that the Qing empire was a distinctively Manchu enterprise. Because place names are labeled exclusively in Chinese, the distinction between China and the Manchu empire is less evident—as on European versions of the maps. Provincial borders (which were also international borders in some cases) are still clearly marked, but the distinction between China proper and the rest of the Manchu empire is less striking. It is these Chinese-language maps and their European derivations, which make China and the other parts of the Qing empire appear as one coterminous entity—when in fact they were distinct administrative and cultural spheres—that have been most influential in determining the shape of the modern Chinese nation-state.

Public Claims vs. Security Concerns

With cartography a delicate balance always persists between the map as an instrument that makes claims—which must be made public to be effective—and the map as conveying highly sensitive information demanding confidentiality. How did the Qing court reconcile the need for international cooperation and recognition in its cartographic endeavors with equally pressing security issues surrounding this sensitive information? Interconnections between the personnel and institutions involved demonstrate the complexity of these competing claims.

These interconnections between individuals, states, and geographical interests in the three countries are seen in the relationship between Antoine Gaubil—a French Jesuit serving in China from 1722–1759 and a member of the Academies of Science of France, Russia, and Britain—and Joseph Nicolas Delisle—a Frenchman serving at the Russian court and a member of the Academy of Sciences of France and Russia. In 1725 they initiated a regular correspondence that lasted for decades. French Jesuits relayed geographical information to the Academy of Sciences in Paris, but also to its counterparts in St. Petersburg and London. As communications between Gaubil and Delisle show, international correspondence was carried out independently, and if not secretly at least

---

72. "Kangxi Map Atlas," maps c.11.d.15 (1721). Pages measure 24″ tall by 15.75″ wide; *Fenfu zhongguo quantu,* 15271.a.20 (1721). Pages are 11″ tall by 5.75″ wide.

with a certain amount of circumspect discretion. Examples of this can be found in the attention to detail regarding how letters and packages should be addressed and to whom entrusted. In a letter of July 1734 Gaubil wrote to Delisle:

> You had thought that the meeting with the Chinese ambassadors would provide a favorable opportunity to write, but for us this is not a secure route. It is better to take advantage of those envoys sent by the Russian court, and it is necessary to be sure that those entrusted with the letters will deliver them into our own hands; otherwise it is dangerous for us.[73]

While undoubtedly the Qing court would have expected its foreign employees to engage in some international correspondence, independent communications with compatriots in the service of foreign courts were not encouraged.

The same was true in earlier decades as well. An edict issued by the Kangxi emperor on the occasion of an imperial audience for Fr. Giovanni Laureati, S.J., on December 3, 1719, is particularly explicit on this point:

> Among you who have come to China, some are good and some bad. As Li Kuo-an [Laureati] is the new Visitor, he must, in the future, first report to the Throne any letter that may come from the West, and must not conceal anything. And should there be among you Westerners those who continue as in past instances to promote difficulties, promiscuously sending letters, they are accordingly subverters of the law and hence are useless to China. In that case, with the exception of those who are skilled in the arts and to be retained in our service, all the rest of the Westerners must be expelled and are under no condition allowed to remain.[74]

Although the Jesuit presence was welcomed by the Kangxi emperor because of the services it could provide at his court, the Qing court was aware of the Jesuits' capability and inclination to share their findings and wished to curtail this eventuality.

It may have been largely for this reason that the Jesuits who worked on the surveys (like all European missionaries after 1706) were required to sign an agreement stating they did not intend to return to their native countries. The Kangxi emperor's description of the undesirability of too much (mis)information circulating was especially colorful.

73. Gaubil, *Correspondance de Pékin*, 371 (letter to Delisle, July 1734).

74. The translation is from Antonio Sisto Rosso, O.F.M., *Apostolic Legations to China of the Eighteenth Century* (South Pasadena: P. D. and Ione Perkins, 1948), 328. The original is found in *Kangxi yu Luoma shijie guanxi wenshu*, document 10.

Hereafter we will permit residence in China to all those who come from the West and will not return there. Residence permission will not be granted to those who come one year expecting to go home the next—because such people are like those who stand outside the main gate and discuss what people are doing inside the house. [75]

The dissemination of certain kinds of information was not in the best interest of the Qing. This may also be the reason why the Jesuit survey-ors' access to the most sensitive frontier areas, and to information about them, was restricted. When they were making the surveys for the maps of the empire they were not allowed to visit certain northern border areas near Russia. The Manchus who were with them, providing assis-tance by order of the Kangxi emperor, made it clear that it would be "useless" and "even dangerous" for them to make that portion of the journey. Undoubtedly these Manchus had orders about which lands the missionaries should and should not enter. Kangxi probably feared that they would transmit any information they found to both France and Russia directly. These lands were visited by other special imperial en-voys, but even as late as 1752 the Jesuits were not allowed access to their findings. Although information on the areas between the Russian and Chinese empires was housed in the Qing court, Gaubil was denied ac-cess to it. Gaubil concluded that he would have to learn about these areas through Russia. [76] Consequently he repeatedly made written re-quests to Delisle for geographical information on the region.

A further reason for the restrictions placed on Jesuits in sensitive frontier regions may have been genuinely related to their safety. Security issues were not just limited to a question of bodily harm, but also to the integrity of the information they had obtained in conducting the surveys and the skills they possessed. John F. Baddeley's discussion of a map known as "Renat I," which was transmitted to Sweden by Renat, a Swedish prisoner of war who worked for seventeen years in the service of Galdan Tseren and his successor—Kalmuk rivals with whom the Qing were at war for many years—throws interesting light on this pos-sibility. Renat indicated that the map was made by Galdan, but whether the leader himself literally made it or had it commissioned is not clear.

75. As quoted in Spence, *Emperor of China*, 82. I am grateful to Pingyi Chu for helping me to locate the original, which appears in *Kangxi yu Luoma*, document 2. See also Lo-shu Fu, *Documen-tary Chronicle of Sino-Western Relations*, 114. On the difficulty of obtaining permission to return to Europe, see Ripa *Memoirs of Father Ripa*, 131–32.

76. Gaubil, *Correspondance de Pékin*, 715–16 (letter to P. Berthier, November 19, 1752.)

Baddeley's careful analysis finds Renat's explanation for the origin of the map plausible. Significantly the map contains detailed information not found on either Chinese or Russian maps of the period. He further speculates that the map may have been commissioned by Galdan from prisoners, including Swedes, who were capable of sophisticated surveying techniques.[77] We do know that in other areas, including artillery manufacture (in which Renat was directly involved) and the establishment of factories for the production of cloth and other commodities, Galdan and his successor put technologies from Europe to good use.[78] In this context, Kangxi may have had legitimate worries about the capture and loss of his own foreign experts, along with their expertise to the other side.[79] If Galdan was indeed using maps made with technology derived from Europe, this simply strengthens the argument that early modern technologies were in no way confined to European use or European culture.

As suggested above, a constant tension existed between the need for secrecy and the need to assert territorial claims internationally. The Qing may simply not have cared to record all the information it had about the northern border areas in the Kangxi atlas—a copy of which had been sent as a gift from Kangxi to Peter the Great in 1721, where Delisle had access to it in St. Petersburg.[80] The gift indicates Kangxi's desire to make his competitor to the north aware both of the international state-of-the-art science the Qing patronized, and of the claims to empire that the atlas conveyed. Yet, it would not do to give away too much.

Aware as he was of European global expansion and the technology making it possible, it is not surprising that the Kangxi emperor wanted his empire mapped. Mapping would serve a number of functions. Most obviously it would allow for better knowledge of the realm and concomitant military advantages in both conquest and subduing revolts. But representing territory cartographically was also one way to lay claim to it. Using scaled maps, easily interpretable by anyone trained in the

77. Baddeley, *Russia, Mongolia, and China,* vol. 1, clxviii.

78. Ibid., clxxvii. On Renat's maps, see also Perdue, "Boundaries, Maps, and Movement," especially 279–81.

79. In 1738 Renat lent a copy of the map known as "Renat 2" to Joseph Nicolas Delisle. "Renat 2" was a copy of a Qing map that the Kalmuks obtained during a battle with the Qing. (Baddeley, *Russia, Mongolia, and China,* vol. 1, clxxviii).

80. Walter Fuchs, *Der Jesuiten-Atlas der Kanghsi-Zeit* (Peiping: Fu Jen University, 1943), 43. See also Foss, "Western Interpretations of China," 250, note 104.

same map idiom, was an effective way to stake out claims of empire to an encroaching Europe; the Kangxi atlas defined what China was territorially to the rest of the early modern world and remained the standard map of China internationally for well over a century. Even today the parameters marked out on the Kangxi atlas continue to influence international opinion on what constitutes China.

CHAPTER THREE

Depicting Peoples

Whereas Chinese cartography has been studied with both implicit and explicit reference to Western cartography, Chinese ethnography prior to the twentieth century has hardly been thought of as ethnography at all. But as with certain types of mapping practices, depiction of peoples became increasingly specialized during the late Ming through mid-Qing and, like European ethnography, had an active role in constructing empires and defining imperial subjects. Just as the specialized practice of cartography emerged and distinguished itself from broader practices of mapmaking generally, so too ethnography gradually emerged as distinct from broader types of descriptions of other peoples.[1] Whereas cartography is based on criteria of precise measurement determined through surveys, ethnography is based on direct observation of peoples. Both in European countries and in Qing China we see over time an increased emphasis on direct observation as opposed to reliance on historical sources alone in the construction of ascribed identities.[2] As with lands surveyed in the cartography of the early modern period, the peoples portrayed were often being studied or observed firsthand for the first time. To illustrate this shift to ethnographic representation and its related uses

---

1. The earliest example of the term's usage in the Oxford English Dictionary dates from 1834. The term was used in Germany, however, already by c. 1790. See G. De Rohan-Csermak, "La première apparition du terme 'ethnologie,'" *Ethnologia Europaea* 1.3 (1967): 170.

2. We may not find the Chinese ethnography of non-Han groups written in the eighteenth century to be particularly "objective" in the attitudes it displays. Nonetheless, methodological standards privileging observation guided those collecting and compiling this information.

in empire building, I begin by describing the findings of scholars who have pursued these questions in the European context.

ETHNOGRAPHY IN EARLY MODERN EUROPE

Alfred Crosby's *The Measure of Reality: Quantification and Western Society, 1250–1600* discusses the early roots of the shift from a qualitative to a quantitative emphasis in Europeans' understanding of their environment. He focuses on the development of, and increasing reliance on, techniques of precise measurement in areas as apparently diverse (but actually interrelated) as the graphic arts, music, and concepts of time and space. He attempts to document the beginnings of a major shift in epistemological outlook in early modern Europe. This same type of gradual change can be observed in descriptions of peoples, in which on-site observation gradually became more prevalent even as a burgeoning interest in natural history arose.[3]

Similarities between the methods of ethnography and of natural history cannot be overlooked. Margaret Hodgen, for example, argues that ethnography distinguishes itself from other, earlier descriptions of foreign peoples through its reliance on direct observation, rejection of non-verifiable information, concern with method, and the urge it reflects both to collect and systematically categorize information.[4] According to Mary Louise Pratt, the development of ethnography was at least partly due to an epistemological shift that had taken place in Europe by the early eighteenth century. She approaches this shift through an examination of European travel writing in conjunction with political and economic expansion beginning c. 1750.

In her 1992 book *Imperial Eyes: Studies in Travel Writing and Transculturation,* Pratt explores the relationship between social, economic, and political forces during the eighteenth century, and the urge to explore new territories and record findings about them. She argues that in northern Europe the emergence of "natural history as a structure

---

3. A collection of review essays on Alfred Crosby's *The Measure of Reality* critiques his assertion that the processes whereby reality was measured and perceived during the early modern period in Europe were unique to Europe and as such responsible for its ascendancy after 1800. See *The American Historical Review* 105.2 (April 2000): Roger Hart, "The Great Explanadum," 486–93; Margaret C. Jacob, "Thinking Unfashionable Thoughts, Asking Unfashionable Questions," 494–500; and Jack A. Goldstone, "Whose Measure of Reality?" 501–8.

4. Margaret T. Hodgen. *Early Anthropology in the Sixteenth and Seventeenth Centuries* (Philadelphia: University of Pennsylvania Press, 1964).

of knowledge" and the simultaneous, and not unrelated, push toward exploration of lands abroad

> register a shift in what can be called European "planetary consciousness," a shift that coincides with many others including the consolidation of bourgeois forms of subjectivity and power, the inauguration of a new territorial phase of capitalism propelled by searches for raw materials, the attempt to extend coastal trade inland, and national imperatives to seize overseas territory in order to prevent its being seized by rival European powers.[5]

She makes the case that these intricately connected forces emergent on the European scene are responsible for a new outlook that fostered a strategic interest in learning and writing about peoples in other lands and cultures.

In order to situate her argument in historical terms, Pratt points to two significant events in the year 1735. The first was the publication of *Systema Naturae (The System of Nature)* by Linnaeus (Carl von Linné), in which he established a classificatory system whereby all plant and animal forms could be categorized as they were discovered. The second event was a major effort to determine the earth's exact shape. Pratt argues that "these two events, and their coincidence, suggest important dimensions of change in European elites' understandings of themselves and their relations to the rest of the globe."[6] This new way of viewing the world was characterized by the confidence and insight that everything within it (starting with the plant world) could be ordered by its classification into a proper system. This shift is fundamental in at least three ways. First, knowledge or information about the material world became worthy of attention; second, there was an optimism that one could understand the natural world, thereby gaining a degree of control over it; third, this undertaking was deemed worthwhile. In short, according to Pratt the privileging of scientific method and inquiry into the natural world were the basis for the emergence of ethnography in Europe. Classificatory schemes were similarly sophisticated in China by the early modern period, and the Qing court, like its European counterparts, was aware of the "global" nature of the earth and the corresponding finite nature of its own territory.

Those who have studied the development of ethnography in the West have shown that certain characteristics distinguish ethnography from

---

5. Mary Louise Pratt, *Imperial Eyes: Travel Writing and Transculturation* (London and New York: Routledge, 1992), 9.
6. Ibid., 15.

earlier types of descriptive writing about other peoples. These same characteristics—precisely those that gained ascendancy along with science's privileging of "objective" knowledge based on direct observation and independent verification, and a new emphasis on humankind and social systems as legitimate subjects of scholarly inquiry—were also present under the Qing dynasty. Furthermore, the development of elaborate classificatory schemes for understanding "others" and their concomitant shortcomings and stereotypes, commonly referred to as "orientalism" after Edward Said's book of the same title, were in no way unique to the "occident."[7]

EARLY WESTERN REPRESENTATIONS OF OTHERS[8]

In the West the father of descriptive writings about other peoples was the Greek author Herodotus of Halicarnassus, who lived and wrote during the fifth century B.C. His *Histories* centers around the war between Persia and Greece that reached its height in 480 B.C., but the content of the work is much broader. In order to ascertain the causes of the war, Herodotus looked much further back into history; his interests were not confined to narrow political events. Like the Chinese Historian Sima Qian, discussed below, he wrote about the customs and practices of many different peoples. He learned about some of these peoples through his travels and others only secondhand. His *Histories* concerns, among other things, over fifty different peoples, including their customs, costume, diet, and marriage patterns.[9]

The amount of space that Herodotus devoted to different groups varied considerably depending on how much was known about them, which was in turn largely determined by geographic proximity and amount of contact between the Greek world and any given group. Groups were constituted, in Herodotus's view, by such shared cultural characteristics as language, religion, customs, dress, and common descent. Many of the topics he explored are still considered of ethnographic interest and value today. Yet the rigor with which he analyzed sources may be another matter, and is difficult to ascertain with complete clarity.

7. Edward Said, *Orientalism* (New York: Vintage Books, 1979).
8. I am indebted to Hodgen's work for much of the information in this section.
9. Herodotus, *The Histories,* edited by Walter Blanco and Jennifer Tolbert Roberts, translated by Walter Blanco. (New York and London: W. W. Norton and Company, 1992).

Although it is hard to analyze the accuracy of his work from our vantage point so distant in time, he himself provides us with some assistance. When he includes information of a fabulous or fantastic nature he alerts his readers that he is recording what has come down to him without altering his sources according to his own interpretations. He sometimes includes contradictory reports, suspending judgment in the interest of recording all the information that he can gather. In such cases he adds his own commentary indicating to his readers that he has not been able to confirm all that he reports, and that he himself is skeptical of its veracity. Not surprisingly, as he reports on peoples living farther and farther away from familiar regions the reports become less believable and more fantastic.

Herodotus's *Histories* was used in the Western world as a reference source on foreign peoples for centuries, but the portions that were later excerpted tended to be the more fantastic sections; the skepticism that Herodotus had expressed about the accuracy of some of the information he had included was not passed on. This is true, for example, of Pliny the Elder's (A.D. 23–79) *Historia Naturalis* as well as the work of Pomponius Mela (first century A.D.) and Solinus (third century A.D.).

During the seventh century Isidore of Seville compiled an encyclopedia entitled *Etymologies,* or "origins." The work was a compilation of excerpts of already existing studies in a wide variety of fields: "arithmetic, geometry, music, astronomy; law and chronology; theology; human anatomy and physiology; zoology; cosmography and physical geography; architecture and surveying; mineralogy, agriculture, and military science." Although not systematically organized, descriptive information on various peoples can be found under such headings as "On Languages, Races, Empires, Warfare, Citizens, Relationships" (book 9), and "On Men and Monsters" (book 11). Clothing is listed in the section "On Ships, Buildings, and Garments" (book 19). Many strange creatures appear in the *Etymologies.* These include nations of giants, pygmies, Cyclopes, dog-faced men, and hermaphrodites. The descriptions are colorful and imaginative, but hardly based on direct observation.[10]

Isidore's seventh-century encyclopedia retained authoritative status until the thirteenth century when another encyclopedia, the *De proprietatibus rerum* compiled by Bartholomaeus Anglicus, appeared in Europe

10. Hodgen, *Early Anthropology,* 55, 57.

and was greeted with widespread popularity that lasted for more than three hundred years. Although six centuries separated Bartholomew's work from that of Isidore, yet he relied on Isidore and other earlier writers (fifth and sixth century) as major sources. Bartholomew also included additional cultural descriptions of peoples closer to hand that were not (apparently) based on older sources. The privileging of direct observation was, however, still far from becoming a priority.

During the late thirteenth and early fourteenth centuries some travelers wrote accounts remarkable for the degree of their foundation in direct observation. These included Marco Polo's (1254?–1324?) account of his travels, and the writings of two Franciscans, Carpini and Rubruck.[11] Carpini's work, entitled *Historia Mongolorum,* is a kind of travelogue/itinerary of Mongolia that includes much analytical description. Rubruck's account deals with much the same geographical area and does so in a similar manner. But the style and content of these writings did not catch on. They were soon largely forgotten in favor of more fanciful, less factually oriented works by such authors as Sir John Mandeville (d. 1372) and Dante Alighieri (1265–1321), the content of which resembled that of the old encyclopedias in its orientation.

Gradually, more firsthand accounts of foreign places were produced. The diaries of Christopher Columbus, although they may look biased and narrow-minded from our vantage point five hundred years after they were written, were indeed unusual and remarkably candid in their observations for their day.[12] Hodgen describes these diaries as containing "realistic, down-to-earth judgments of the Caribs and their culture" and cites them as constituting "almost unique departures in ethnological attitude and method of inquiry." Columbus also contrasted his observations with the kind of material found in the literature about foreign lands available in Europe at the time.[13] Although some of his observations were based on his own personal experience, other parts were based on hearsay. He writes, for example, of reports of people with one eye in the middle of their forehead or the face of a dog.[14] One letter describes people born with tails, and islands where no men, but only women,

11. The lack of certain details in Polo's work—mention of foot binding, for example—has led Frances Wood to question whether Marco Polo actually reached China on his travels. Frances Wood, *Did Marco Polo Go to China?* (Boulder, Colo.: Westview Press, 1996).

12. Oliver Dunn and James E. Kelley, Jr., trans., *The "Diario" of Christopher Columbus's First Voyage to America, 1492–1493* (Norman, Okla., and London: University of Oklahoma Press, 1989).

13. Hodgen, *Early Anthropology,* 20.

14. Dunn and Kelley, *The "Diario,"* 167, 177.

lived, and reports he had heard of cannibalism.[15] The above examples attest that at this time when direct observations were recorded they were placed in addition to and alongside other (often contradictory) information passed down through the ages[16]—as in the Ming and Qing encyclopedias, the *San cai tu hui* and *Gujin tushu jicheng,* discussed below.

Columbus's first report on his voyages (1493) was published soon after his return to Europe. Before a year had passed it was available in Spanish, Italian, German, and Latin editions. These initial editions did not contain illustrations, but before long various editions were supplemented by woodcuts, and versified editions in Italian also appeared.[17] These early illustrations were based on observation only to the degree possible by relying on Columbus's transmissions. The clothing—what little there was—more nearly resembled that of people of medieval European legend than what would actually have been encountered in the New World. Additionally the depictions of spears also reveals attention to, but misunderstanding of, the text. Finally, some of the figures shown are bearded, or have wavy hair—physical characteristics not found among Native American populations.[18] Thus these earliest drawings of the New World were based on written descriptions rather than on direct observations of models by the artist. They also remind us that standards for "objectivity" and accuracy in portrayals of other peoples emerged only gradually and were not yet fully prioritized in reports on early explorations.

EARLY CHINESE REPRESENTATIONS OF OTHERS

The earliest Chinese book to describe peoples from other lands is known as the "Classic of Mountains and Seas," or *Shan hai jing.* The origins of the work are not entirely clear. We do know, however, that it

15. Alden T. Vaughan, "People of Wonder: England Encounters the New World's Natives," 11–23, in *New World of Wonders: European Images of the Americas, 1492–1700,* ed. Rachel Doggett with Monique Hulvey and Julie Ainsworth (Washington, D.C.: The Folger Shakespeare Library), 14.

16. Sebastian Münster's *Cosmographia: Beschreibung aller länder* (Basel, 1544) is also a good example of a hodgepodge of information that to us seems to lack logical arrangement (Hodgen, *Early Anthropology,* 144). His description of the Turks is longer than any other entry. This undoubtedly bears a correlation to the Europeans' continuing preoccupation with and fear of the Turks (and Mongols), a product of their near invasion of Europe in the sixteenth century. See also Hodgen, *Early Anthropology,* 147–48.

17. Sturtevant, William C. "The Sources for European Imagery of Native Americans," 25–33, in Doggett, Hulvey, and Ainsworth, *New World of Wonders,* 25.

18. Ibid., 25.

was written by numerous contributors at various times, and that it had been completed by the time of the Han dynasty (206 B.C.–221 A.D.). Primarily devoted to physical geography, it also contains descriptions and illustrations of the residents of foreign lands. Many of the creatures represented in the drawings seem fantastic; they may be part human and part animal, or have unusual numbers and combinations of body parts. The illustrations, which were added at a later date, derive from a literal reading of the texts. For this reason they may not in fact represent the images that the authors intended to convey.[19]

The Han dynasty historian Sima Qian included a variety of ethnographic information in his *Shi ji,* or *Records of the Historian.*[20] This work represented a new departure in Chinese historical writing both in scope and in organization (as such it can be compared to Pei Xiu's roughly contemporaneous grid maps of the country). Sima Qian approached his sources with a critical attitude. Like Herodotus, he is known for his role as a recorder of information who consciously stood apart from what he recorded, commenting on it from aside, but not altering the contents of what he had heard by word of mouth or learned through written sources. Accounts of foreign lands and peoples are found in the biographical section *(lie zhuan)* of the work.

Following the Han period, works containing representations of foreigners were not uncommon. Records of illustrations of tributaries were discussed above, in chapter 1. In the year 661, a prince from Tocharistan submitted to the emperor a description of his lands and peoples under the title *Xi yu tu ji* (illustrations/maps and records of western regions). According to Ma Duanlin (1254?–1323), the volume accompanied a request made by the prince for Chinese administration of certain regions of his domain.[21] Yan Liben, a Tang dynasty (618–906) court painter, is renowned for his illustrations of foreign tribute bearers. Wang Shang, who lived during the Five Dynasties period (907–960), is also known for having made illustrations of tribute bearers.[22] The Tang encyclope-

---

19. The first illustrated edition may have appeared during the Jin dynasty (265–420 A.D.). For a comprehensive study of the *Shan hai jing,* see Remi Mathieu, *Étude sur la mythologie et l'ethnologie de la Chine ancienne* (Paris: Collège de France, Institut des Hautes Études Chinoises, 1983).

20. For a comparison of Sima Qian and Herodotus, see Teng Ssu-yü, *Ssu-ma Ch'ien yü Hsi-lo-to-te chih pi chiao (Szu-ma Ch'ien and Herodotus), Bulletin of the Institute of History and Philology Academia Sinica,* vol. 27, *Studies Presented to Hu Shih on His Sixty-fifth Birthday* (Taiwan: Academia Sinica, 1957), and *Herodotus and Ssu-ma Ch'ien: Two Fathers of History* (Rome: Ismeo, 1961).

21. Hirth, "Über die chinesischen Quellen," 232.

22. Friedrich Hirth, *Über fremde Einflüsse in der chinesischen Kunst* (Munich und Leipzig: G. Hirth, 1896), 50. See also Hirth, "Über die chinesischen Quellen," 237. *Xuan he hua pu,* 3, 1a–b.

dia *Tong dian* (eighth century) lists several works about foreign customs that may or may not have been illustrated. This includes *Zhu fan fengsu ji* (a Record of the Customs of All the Western Tribes), *Tujue suo chu fengsu shi* (Affairs and Customs Emanating from the Turks) and *Waiguo tu* (Illustrations/Maps of Foreign Countries).[23] The character *tu*, often translated as illustration, can also mean plan or map and thus cannot always be equated with the English term "illustration." Travel accounts written by Buddhist monks who journeyed to India were also produced during the Tang.

The Song Dynasty (960–1279) existed under the ever-present threat of invasion from the north. Successful foreign policy depended on reliable sources of information about the outside world.[24] The *Zhu fan zhi* (A Record of All Barbarians) compiled by Zhao Rugua in 1225, for example, unlike earlier texts, contains information on countries that did not pay tribute to China, and also pays more attention to both cultural and political characteristics of a variety of states.[25] Painting too, much of which was relatively realistic in its depictions during this period, sometimes took foreign countries and customs as a theme. The *Xuan he hua pu*, a catalogue of paintings published in 1120 A.D., contains the subject heading *fan zu*, or "Barbarian Tribes," which includes

> pictures of nomad life beyond the frontier, Turkish, Tibetan, and other types foreign to China; horsemen in foreign costume, caravans and hunting scenes, tribute bearing missions, and Chinese princesses going abroad with their escorts; strange customs, unknown animals, and other remarkable productions of alien lands.[26]

Tribute illustrations also continued to be painted during the Song.[27]

With regard to the depictions of minority peoples within China an interesting note appears in *Zhu fan zhi*. It says that in 1174 the prefect of

23. Hirth, "Über die chinesischen Quellen," 228.

24. It may be no coincidence that it was also during the Song that we see accurate to-scale maps such as the *Yu ji tu*.

25. Michel Cartier, "Barbarians through Chinese Eyes: The Emergence of an Anthropological Approach to Ethnic Differences," *Comparative Civilization Review* 6 (1981): 10. For a translation of Zhao's work, see William W. Rockhill and Friedrich Hirth trans., *Chao Ju-Kua: His Work on the Chinese and Arab Trade in the Twelfth and Thirteenth Centuries, Entitled Chu-fan chih* (New York: Paragon Book Reprint Corp., 1966).

26. Stephen W. Bushell, *Chinese Art*, vol. 2 (London, 1909), 138.

27. Jaeger, "Über chinesische Miaotse-Albums" (1916), 267. The imperial burial grounds of the Song dynasty included statuary that realistically depicted foreign envoys. Ann Paludan, "Foreign Influences on Northern Song Statuary" (paper presented at the twenty-first annual meeting of the Mid-Atlantic Region Association for Asian Studies Conference, West Chester University, West Chester, Pennsylvania, October 30–November 1, 1992).

Qiong department (now Hainan province) had likenesses made of the outer manifestations and clothing of the original inhabitants, known as Li, to present to the governor of the province[28] when they submitted to Chinese authority.[29] The format and purpose of this work seems to have been similar to the illustrations of tributaries and even the later eighteenth-century Miao Albums.

The Chinese Ming dynasty encyclopedia, *San cai tu hui*, was compiled by a scholar named Wang Qi (1565–1614), and published circa 1610 in Nanking.[30] It is divided into fourteen sections, two of which are most relevant to our purposes here. They are the sections on geography *(dili)* and personages *(renwu)*. The geography section includes maps of all the provinces of the Ming, accompanied by a textual description of the history of the area mapped, its geographical characteristics, and the administrative units into which it had been divided. The map of Guizhou province is reproduced below (map 2). The geography section is also comprised of illustrations and explanations of famous places, including Red Cliff, famous as the site of a pivotal battle during the Three Kingdoms period (220–280) that was later immortalized in the Ming dynasty historical novel *Romance of the Three Kingdoms*. Chapters 12 to 15, headed *renwu*, or personages, display a variety of peoples and costumes. Most if not all of the illustrations are based on earlier drawings, some of which are no longer extant. Many of the illustrations in the section on foreign peoples seem fantastic, although they do not all defy belief. Figure 16 reproduces several illustrations from the Ming encyclopedia that were probably taken directly from the Classic of Mountains and Seas. The first one shows a "man with long arms" and a "man from the country of long legs." The second and apparently more realistic set of illustrations is of men from "Chao lu dong" and "Hui hui" country.[31] Although one shouldn't push the parallels too far, the Ming departure from Song objectivity and the inclusion of reproductions from the *Shan hai jing* in the *San cai tu hui* remind one of Mandeville's overshadowing of Carpini's and Rubruck's work. Like Münster's *Cosmographia*, this Ming encyclopedia contained both material based on direct observation and nonverifiable descriptions of strange creatures and peoples from afar.

28. Hainan did not become an independent province until the 1980s. From Ming times it had been under the jurisdiction of Guangdong.

29. Jaeger, "Über chinesische Miaotse-Albums" (1916), 267.

30. John A. Goodall, *Heaven and Earth: 120 Album Leaves from a Ming Encyclopedia, San-ts'ai t'u-hui, 1610* (London: Lund Humphries, 1979), 11.

31. Probably refers to a Turkish or Muslim area.

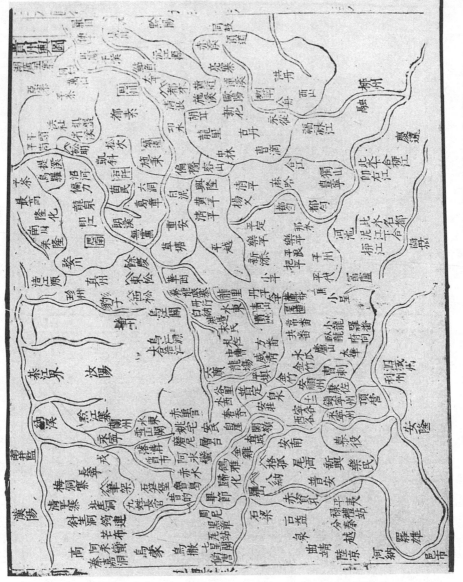

Map 2 Guizhou Province. *San cai tu hui.* Reproduced with permission from Ch'eng Wen Publishing Co, Taipei.

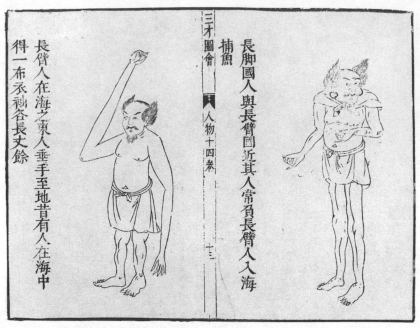

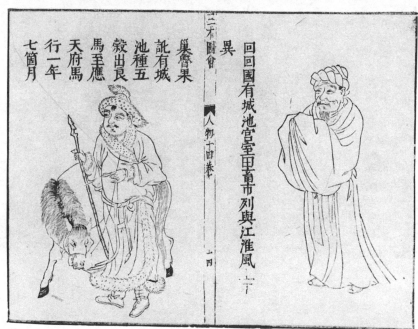

Figure 16 Men with Long Arms and Long Legs, and Foreigners. *San cai tu hui*. Reproduced with permission from Ch'eng Wen Publishing Company, Taipei.

Additional Ming works also showed an interest in other peoples. The *Yi yu tu zhi* (Pictures and Descriptions of Strange Regions) was printed from woodblocks and contains 168 annotated illustrations of various places including foreign countries.[32] The Report of the Librarian of Congress for 1935 contains a reference to *Dong yi kaolüe* (A Study of Eastern Barbarians), by Mao Ruizheng, from the early seventeenth century.[33] Another early seventeenth-century Ming work, also in the Library of Congress, entitled *Wanyong zhengzong buqiuren quanbian* (The "Ask No questions" Complete Handbook for General Use), contains a section on the customs of foreign countries *(Zhu yi za zhi)*. The illustrations of this section are the same as those appearing in the *Yi yu tu zhi*.[34] Another Ming work, the *Xianbin lu* (A Record of All Guests), printed c. 1590–91, also contains information about foreign countries. This work describes more than one hundred countries, lists about "850 clan names of foreign peoples," and also indicates the sources on which it drew. A record of foreign embassies to China is also included.[35] The vast majority of the twenty-four dynastic histories, standard Chinese historiography par excellence, also include a special section devoted to foreign peoples.[36]

The Ming imperial government also showed interest in the languages of border groups. J. Edkins, in an article in the *Chinese Recorder*, stated that he had woodcuts of vocabularies in two tribal languages that had been used at an imperial language school in Peking during the Ming dynasty.[37] This may be seen as evidence of early *kaozheng*, or evidential learning, scholarship.

The early Qing encyclopedia, *Gujin tushu jicheng*, contains what may seem to us in retrospect different types of information appearing to de-

32. Arthur. C. Moule, "An Introduction to the *I yü t'u chih* or 'Pictures and Descriptions of Strange Nations' in the Wade Collection at Cambridge," *T'oung Pao* 26 (1930): 180.

33. Arthur Hummel, "Eastern Barbarians," *Annual Report of the Librarian of Congress for the Fiscal Year Ending June 30, 1935* (Washington, D.C.: U.S. Government Printing Office, 1936), 192–93.

34. Arthur Hummel, "A Ming Encyclopedia . . . ," *Annual Report of the Librarian of Congress for the Fiscal Year Ending June 30, 1940* (Washington, D.C.: U.S. Government Printing Office, 1941), 650–51. The translations of the titles are Hummel's.

35. Arthur Hummel, "A View of Foreign Countries in the Ming," in ibid., 167.

36. See Cartier, "Barbarians through Chinese Eyes," 1–14. Cartier's article looks at the ethnographic sections of the twenty-four histories and categorizes what is discussed in them. He argues that Chinese writings about others do gradually become more "ethnographic" over time. See especially pp. 3, 4, and 8.

37. J. Edkins, "The Miau Tsi," *Chinese Recorder and Missionary Journal* 2 (August 1870): 76.

rive from opposing methodological principles. The personages *(renwu)* section contains illustrations borrowed directly from the *San cai tu hui,* and the earlier Classic of Mountains and Seas, of creatures whose appearance is bizarre and whose actual existence implausible (figure 17).[38] Yet in the geography *(dili)* section of the same work, which deals with political divisions of China *(zhi fang)*, we see the advent of ethnographic information about the peoples who lived in the regions under discussion—information that could be ascertained through firsthand observation and independently verified. The texts describe plausibly the customs of the residents. The inhabitants are listed according to where they lived. The type of content is similar to what is found in the Guizhou gazetteers and may even have been excerpted from them. There are no accompanying illustrations. Thus in the *Gujin tushu jicheng,* as in early European works on the new world, we see a combination of types of information: some derived from older literary sources, some that seems to have been based on relatively contemporary direct observation.

As with physical geography, dynastic China had seen an interest in human geography for millennia, and early sources that have come down to us show varying degrees of concern with the possibility for verification of the information they contained. In both branches of geography we see more value placed on precise measurement and direct observation during the Han and the Song, with a drop-off in this kind of method during other periods. However, up through the beginning of the Qing dynasty, even when direct observation is used to gather information, concurrent reliance on previous nonverifiable literary sources is apparent. The *Gujin tushu jicheng,* for example, included content from earlier well-known literary sources without independent confirmation of its accuracy.

The above comparative historical overview shows that political uses of representation, even in efforts to build empire, are not limited to the early modern period. However, just as early modern cartography distinguished itself from mapmaking more generally, we can draw a distinction between what can be termed "ethnography" and these earlier representations. The emergence of just such an early modern ethnography in the southwestern reaches of the Qing is described in chapter 5. Ultimately my objective is not to show how visual representation under the

---

38. The texts accompanying the illustrations begin with the statement "According to the *Shan hai jing. . . .*"

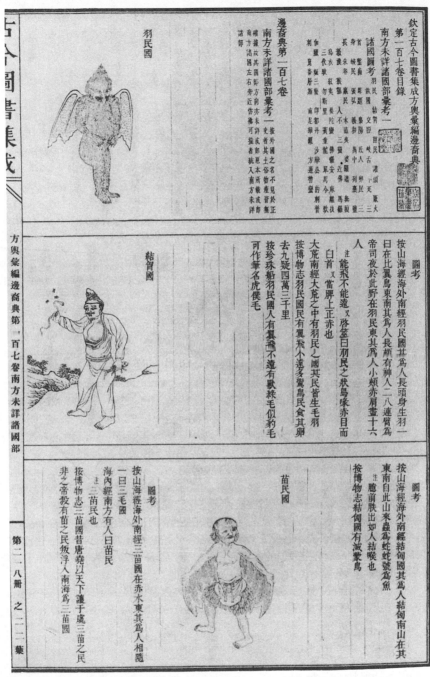

Figure 17 Illustrations Based on Texts of the *Shan hai jing,* or Classic of Mountains and Seas. *Gujin tushu jicheng.*

Qing was different from earlier periods in Chinese history, but rather to demonstrate Qing interconnectedness with the concerns and technologies of the larger early modern world.

ETHNOGRAPHY AND COLONIALISM

Scholarship on nineteenth- and early twentieth-century ethnography has argued for its close association with the colonial enterprise. Talal Asad was one of the first to make this argument. His 1973 edited volume *Anthropology and the Colonial Encounter* [39] deals with prewar social anthropology in Britain. It was among the earliest works to attempt to articulate the power relationship that can exist between those who carry out ethnographic research and those who are in a position to exploit the information thus obtained. He sees anthropology as

> one . . . discipline among several (orientalism, indology, sinology, etc.) [that] . . . are rooted in that complex historical encounter between the West and the Third World which commenced about the 16th century: when capitalist Europe began to emerge out of feudal Christendom; when the conquistadors who expelled the last of the Arabs from Christian Spain went on to colonise the New World and also to bring about the direct confrontation of "civilised" Europe with "savage" and "barbaric" peoples; when the Atlantic maritime states, by dominating the world's major seaways, inaugurated "the Vasco Da Gama epoch of Asian History"; when the conceptual revolution of modern science and technology helped consolidate Europe's world hegemony.[40]

Asad is correct, I think, in identifying the potential for an exploitative relationship between ethnographer and subject. His framework of "Western" and "Third World" is, however, too narrow—the product of a dichotomy that does not allow a place for recognition of the same phenomenon in other contexts. The central issue is, rather, how centers of power with a monopoly on the production and dissemination of knowledge define peripheral groups and attempt in one way or another to dominate them. The struggle for control is not only a product of "Western" hegemony.

Michael Hechter's 1975 work on "internal colonialism"[41] centers geographically on England's relationship with its Celtic territories,

39. Talal Asad, ed., *Anthropology and the Colonial Encounter* (London: Ithaca Press, 1973).
40. Talal Asad, "Two European Images of Non-European Rule," in Asad, *Colonial Encounter*, 103–4.
41. Michael Hechter, *Internal Colonialism: The Celtic Fringe in British National Development, 1536–1966* (Berkeley and Los Angeles: University of California Press, 1975).

demonstrating how the latter were forced into economically and politically dependent positions that were mutually determined.[42] His findings are interesting as a comparative case to China not only because of similar processes that took place, but also because of the geographic proximity of the territories gradually incorporated into the empire.

His work provides another demonstration of how ethnographic description can serve in contexts of unequal power relations to assist in circumscribing and limiting indigenous peoples. Denigrating the culture of those who are co-opted into a subservient position is part of this process. In order to "advance" in the newly imposed dominant cultural and political order, locals must shed their old ways (that are to some extent newly created by those defining them) in order to take on new habits and ways of life acceptable in the eyes of those who control the paths toward upward mobility, whether political, social, or economic.[43] Hechter's findings, while hardly startling now, were groundbreaking in their day. By describing a process of internal colonization, he argued that distance is not a major factor in the exploitation of additional territory or the co-opting of neighboring peoples into an economically and politically subordinate relationship.

On a related note, Pratt remarks on similarities in content and tone between European travel writing describing regions outside Europe and those closer to home. She explains the parallel by stating that similar power dynamics are likely to be found in each case. Northern Europe attempted to dominate and subsume Mediterranean areas in much the same way it did territories farther abroad.[44] According to Pratt, ethnographic representation is not simply a response to foreignness or to the distance of the locale portrayed, but is rather the product of a power dynamic. We cannot say that the Qing was not a colonial power simply because its expansion did not involve lands overseas.

Bernard Cohn has written on the relationship of colonization to the practice of anthropology in India and has come to the conclusion that "[i]n a very real way the subject matter of anthropology has been the study of the colonized."[45] In an article on British census taking in India, Cohn shows how the gathering of information about a group of people

---

42. Ibid., 81.
43. Ibid., 73.
44. Pratt, *Imperial Eyes*, 10.
45. Bernard S. Cohn, "The Census, Social Structure and Objectification in South Asia," in his book *An Anthropologist among the Historians and Other Essays* (Delhi: Oxford University Press, 1987), 224.

to be ruled was carried out; how the information thus acquired was used to differentiate different subgroups (specifically caste groups); and how the resulting categorization of these different groups of ruled peoples in turn influenced the ways they behaved and how they thought about themselves. The parallels between these British census materials made of India and the Qing Miao albums discussed below are considerable. Both projects allowed for improved understanding of and control over their subjects.[46]

The political implications of representation generally, and ethnographic representation in particular, have been thoroughly critiqued.[47] Jonathan Friedman has described the dynamic inherent in ethnographic representation as follows:

> As ethnographic description is the practice of writing the Other for us, here at home, it precludes, by definition, the voice and the pen of the Other. Ethnography, thus, embodies the authority to represent and, by logical implication, the authority to maintain the Other in silence. Now this is a serious political act since it identifies the Other for us. It also, ultimately, through colonial and post-colonial apparatuses, returns that identity to the Other so that it becomes, by hook or by crook, the latter's own identity. So the issue is not merely a disciplinary one, but strikes at the heart of the general relation between power and representation.[48]

Although the context of Friedman's comment is specifically related to ethnographic writing in the discipline of anthropology during the second half of the nineteenth century, his argument also applies to the problem of representation more generally. In this respect, as in others described above, ethnographic writing of the West—the object of Friedman's attentions—and that of eighteenth-century China, share fundamental similarities. In essence those who do the depicting define the peoples described. Dominant groups or powers thus restrict minorities by speaking both about and for them, circumscribing their rights or potential to define themselves. In the colonial context the goal of the

46. For more on the relationship of ethnography and British expansion overseas, see George Stocking's work.

47. For critiques of representation generally, see Michel Foucault, *The Order of Things: An Archaeology of the Human Sciences* (New York: Vintage Books, 1973) and Said, *Orientalism*. For critiques of ethnographic representation, see James Clifford, *The Predicament of Culture: Twentieth-Century Ethnography, Literature, and Art* (Cambridge and London: Harvard University Press, 1988), and George E. Marcus and Michael M. J. Fischer, *Anthropology as Cultural Critique: An Experimental Moment in the Human Sciences* (Chicago: University of Chicago Press, 1986).

48. Jonathan Friedman, "Narcissism, Roots and Postmodernity: The Construction of Selfhood in the Global Crisis," in *Modernity and Identity*, ed. Scott Lash and Jonathan Friedman (Oxford, U.K., and Cambridge, Mass.: Basil Blackwell, 1992), 332.

colonizer, or imperial authority, is precisely and unabashedly to learn about, or rather construct, the identity of those to be ruled. Such knowledge simplifies the task of governance.

The fundamental problem of ethnographic writing has been its nature as the product of an unequal relationship in which representatives of a culturally, politically, or militarily (sometimes all three) dominant culture have attempted to define other peoples.[49] The inequality in the relationship is intensified when the ethnographer intrudes on the territory of those described, or uses technologically sophisticated means unavailable to those portrayed to create and perpetuate images that will define them as subjects. Although the peoples described may have their own opinions about their depictors, they generally have comparatively limited power to propagate and disseminate their own views, especially in frames of reference that would be persuasive to those attempting to exercise control.

Too often unrecognized is the fact that the politics of representation encapsulated in the idea of "orientalism" is not simply a feature of Western modernity, but of the colonial encounter itself, wherever colonial relations are played out. This capacity or inclination to "orientalize" is not unique to the Western world.[50] Techniques of representation grounded in measurement and direct observation, while fundamentally related to our definition of the early modern period, were not unique to early modern Europe.

The attempt to acquire a measure of control over others through the power to represent them is not unique to Europe, or new with the advent of modern ethnography. We saw evidence of this kind of political

49. One of the preoccupations of present-day ethnography is how to avoid these problems. Some ways to work toward a solution are the inclusion of multiple voices—specifically those of the ethnographic subjects—and avoidance of government sponsorship of or involvement in ethnographic projects. See James Clifford and George E. Marcus, eds., *Writing Culture*, especially Paul Rabinow's essay "Representations Are Social Facts: Modernity and Post-Modernity in Anthropology" (Berkeley: University of California Press, 1986), 234–61.

50. On non-Western "orientalism," see Louisa Schein, "Gender and Internal Orientalism in China," *Modern China* 23.1 (January 1997): 69–98; and Stephen Tanaka, *Japan's Orient: Rendering Pasts into History* (Berkeley, Los Angeles, and London: University of California Press, 1993). For Japanese depictions of others in the early modern period, see Ronald P. Toby, "The Race to Classify," *Anthropology Newsletter* 39.4 (April 1998): 55–56, and "The Indianness of Iberia and Changing Japanese Iconographies of Other," in *Implicit Understandings: Observing, Reporting, and Reflecting on the Encounters between Europeans and Other Peoples in the Early Modern Era*, ed. Stuart B. Schwartz (Cambridge: Cambridge University Press, 1994), 323–51. See also Jan van Bremen and Akitoshi Shimizu, eds., *Anthropology and Colonialism in Asia and Oceania* (Richmond, Surrey: Curzon Press, 1999), especially the essay by Margarita Winkel on Japan's colonization of its northern territory.

manipulation of images in the Tang dynasty paintings of illustrations of meetings with kings, and in early tribute illustrations. However, as with advances in eighteenth-century cartography, early modern ethnography has contributed to the systematization of this kind of representation through an increasingly rigorous methodology that privileges observation and schemes for classification. The Qing colonial enterprise in the southwest is a case in point. The chapters that follow document this process in Guizhou province, where the development of ethnographic writing during the seventeenth and eighteenth centuries reflects departures from earlier Chinese depictions of others, and in many ways parallels the development of ethnography in early modern Europe. The texts about Guizhou's populations display evidence of increasing reliance on direct observation to gather information, the elaboration of a system of classifying and naming different groups, and a concern with social institutions, as well as evidence of the collection of this kind of material for its utility to the state in the expansion of its frontiers. As we shall see, the Miao albums are also unique because they constitute a genre that is devoted entirely to the representation and description of non-Han peoples. But it is especially in the close association between collecting ethnographic information and the colonial enterprise that eighteenth-century ethnography of Guizhou province resembles European ethnography of the early modern period.

CHAPTER FOUR

Bringing Guizhou into the Empire

T he second half of this book closely examines the relationship between colonization and the development of ethnographic representation in southwest China, primarily Guizhou province, during the seventeenth and eighteenth centuries. Many of the same processes that gave birth to ethnographic writing in Europe were operative under the Qing as well.

Although Guizhou was named as a province during the Ming dynasty in 1413, the central government had little control over the region until the mid-Qing (see map 3).[1] Imperial penetration of the province was a lengthy process that required becoming familiar with its territory, and demystifying its peoples. In addition to these means, repeated military campaigns aimed to incorporate the remote and mountainous province more fully into the empire. These efforts met with varying degrees of success. Rebellion was a recurring theme in the history of the area throughout the eighteenth and nineteenth centuries. Even today some areas of Guizhou inhabited largely by non-Han groups are governed as autonomous regions.

THE TOPOGRAPHY OF GUIZHOU

Between 1715 and 1716 a team of surveyors, including the Jesuit fathers Régis and Fridelli, visited Guizhou as part of the Kangxi emperor's plan

1. See, J. E. Spenser, "Kueichou: An Internal Chinese Colony," *Pacific Affairs* 13 (1940): 162–72; Lombard-Salmon, *Un exemple d'acculturation chinoise*; James Z. Lee, *The Political Economy of a Frontier: Southwest China, 1250–1850* (Cambridge, Mass.: Harvard University Press, 2000).

to map the Qing empire. The observations of the fathers, as recorded by Jean Baptiste Du Halde, give us a sense of the province at that time:

> It is one of the smallest provinces of China, situated between Huguang,[2] Sichuan, Yunnan, and Guangxi. . . . It is full of inaccessible mountains, which accounts for the fact that a portion of the province is inhabited by people who have never submitted themselves to the emperor and who live in complete independence of the laws of the empire. . . . In accordance with the imperial plans to populate the province, colonies of Chinese have often been sent, sometimes even governors with their entire families.
>
> There is a large quantity of forts and places of war where numerous garrisons are supported; the tribute that is drawn from the province is not sufficient for their subsistence. For that reason the court is obliged to sustain them. Each year it sends assistance.
>
> The mountains contain mines of gold, silver, and mercury. Copper used for minting coin all over the country also comes in part from this province.
>
> Between the mountains one sees valleys which are pleasing and relatively fertile, especially near the rivers. Commodities are available at a good price, but not with the same abundance in which they are found elsewhere and which could be found here were the land better cultivated. . . .[3]

These contemporary observations highlight certain aspects of life in the province often overlooked by Chinese sources. Most striking is the frank way in which they speak of Guizhou as a colonial space. Furthermore, despite the considerable resources poured into the province each year, large areas remained outside the control of the government. This lack of control is graphically reflected on the 1721 map resulting from the Jesuit surveys in which considerable blank space bordering Guizhou is marked off as Sheng (wild) Miao territory (see maps 4 and 5).[4]

In the eighteenth century there were four main routes traversing Guizhou, all of which led to Yunnan. They were called the east road *(dong lu)*, west road *(xi lu)*, middle road *(jian lu)*, and finally the Guangxi road (see map 6). The middle road dates from at least the second century A.D. and may be the oldest. The western route, used during the Tang Dynasty (618–906), was abandoned for official uses in the

2. Huguang comprised the rough equivalent of what are now Guangxi and Guangdong provinces.

3. Du Halde, *Description . . . de la Chine . . .* (Paris: P. G. Lemercier, 1735), vol. 1, 255.

4. *Sheng* and *shu*, which can be translated alternately as "wild" and "tame," "unpacified" and "pacified," or even "raw" and "cooked," are often used as modifiers in classical Chinese documents to classify non-Han peoples into two broad categories of generalization. As taxonomies became more complex the term *sheng miao* continued to be used in a general sense for unpacified non-Han in the southwest. *Sheng Miao* also came to refer to a specific group of Miao within Guizhou province.

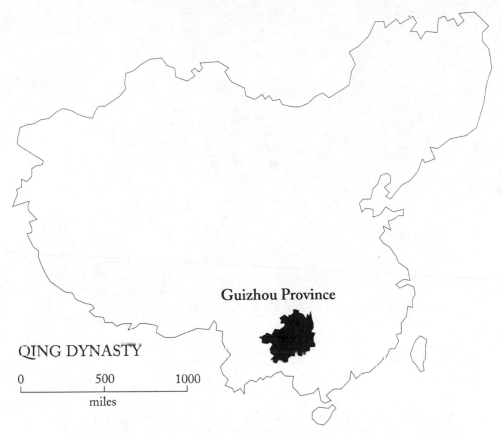

Map 3 The Qing, c. 1820, Showing Guizhou Province. Cartography Laboratory, Department of Anthropology, University of Illinois at Chicago.

beginning of the Ming when it was taken over by an Yi (Luoluo) chieftain named She Xiang. Under the Qing it came back into the hands of the central government. The eastern road, although in existence earlier, was set up with relay stations during the Ming. From that time the imperial administration guarded it as the principal artery for travel through Guizhou to Yunnan. The fourth major road led south toward Guangxi and then east to Guangdong.[5]

In addition to these major official roads, networks of smaller routes and paths also crisscrossed the province. These networks tended on

5. Lombard-Salmon, *Un exemple d'acculturation chinoise*, 93–95. The famous general Zhuge Liang (181–243 A.D.) is said to have used the middle road in his expedition against the southwestern barbarians of his times.

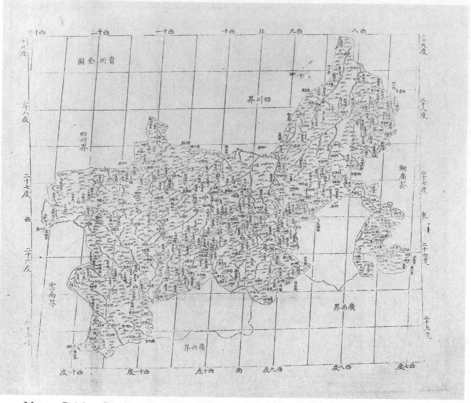

Map 4 Guizhou Province, Reproduced from the 1721 Kangxi Atlas. Note the blank territory marked out to the south of the province. A close look at the map of Guangxi province from the same source indicates this is Sheng (wild) Miao territory. The British Library.

the whole to be much more extensive in the north than in the south. Lombard-Salmon attributes this to the strategic importance of certain areas in the north (for example, in defending against sometimes hostile indigenous populations), to economic and administrative links with Sichuan and Hunan, and finally to the exploitation of rich mineral resources in northwestern Guizhou. The sparsity of roads in the south may be due to the wet and warm climate, which those unaccustomed to it found extremely trying. The local non-Han populations also had their own routes, which were not well known or used by the Han, except perhaps by merchants who visited indigenous villages to sell their goods.[6]

6. Ibid., 95–97.

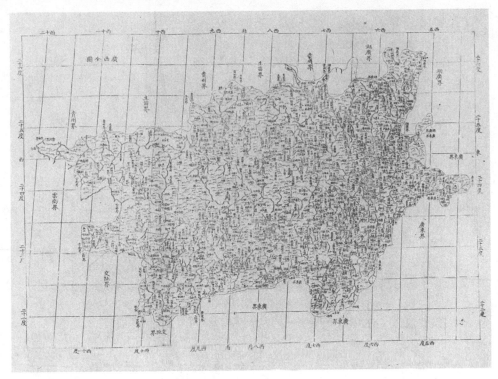

Map 5 Guangxi Province, Reproduced from the 1721 Kangxi Atlas. The area north of the province is marked alternately "border with Guizhou" and "border with Sheng (wild) Miao [territory]." The British Library.

Due to its topographic inhospitality, until the population explosion of the eighteenth century Guizhou was largely protected from intrusion by outsiders unaccustomed to the harshness of life dictated by the environment. Since its terrain was unsuitable for the cultivation of rice, a mainstay in most of south China, Guizhou's diet and related aspects of its culture were alien even to most rural Han Chinese.

Guizhou's Non-Han Populations

As noted above, until at least the mid–nineteenth century, the majority of Guizhou's inhabitants were not Han Chinese. The province is still populated by many "minority nationalities" today. Who are these peoples? It may be helpful to describe the seven main ethnic groups in

Guizhou Province

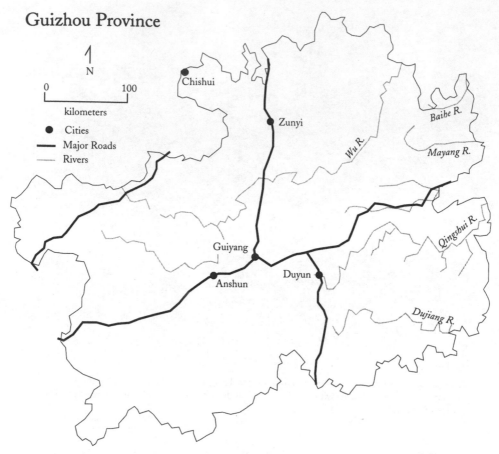

↑
N

0   100

kilometers

●   Cities
━━   Major Roads
───   Rivers

Chishui

Zunyi

Wu R.

Baihe R.

Mayang R.

Qingshui R.

Guiyang

Duyun

Anshun

Dujiang R.

Map 6 Qing Guizhou, c. 1820, Showing Main Roads and Rivers. Cartography Laboratory, Department of Anthropology, University of Illinois at Chicago.

the region according to current ethnolinguistic criterion and what we know of their historical background.

Miao

The largest group is known to their Chinese neighbors as the Miao. To clarify, the term *miao*, like many Chinese characters, has several meanings. The most prevalent meaning of the term is "sprout," as in young seedling. Two secondary usages of the word apply to non-Han groups in southwest China. The more narrow usage denotes a specific ethnic group still found in southwest China. The broader meaning of *miao*, as used in the titles to the Miao albums, refers to non-Han peoples of south China generally.[7]

Today Miao, in the narrow sense of the term, are found in North Vietnam, Laos, northern Thailand, China's southern provinces of Guizhou, Yunnan, Sichuan, and Guangxi, and various cities in the United States.[8] Those outside of China are commonly referred to as Hmong, a term that more closely reflects the group's own pronunciation of its name than does the Chinese transliteration "Miao." Miao diaspora outside China dates mostly from the nineteenth century and later.[9] There are several Miao dialects and even some mutually unintelligible Miao languages. Linguistically they have been classified as Mon-Khmer (Austro-Asiatic), Tai,[10] and Sinitic.[11]

Zhongjia

Who precisely the Zhongjia were, and whence their name derives, is uncertain; sources are sometimes contradictory. According to Lombard-Salmon, the appellation seems to have come into use during the Ming, and different subgroups within the Zhongjia were distinguished during the Qing. The same group was also sometimes referred to as Yiren.[12] According to Robert Ramsey, "[T]hey are not clearly distinguishable from the Zhuang either linguistically or culturally, and they are almost

7. Herold J. Wiens, *China's March toward the Tropics* (Hamden, Conn.: Shoe String Press, 1954), 35. The term *miao man* also appears with some frequency in the titles of the Miao albums. It is a general term that denotes all non-Han peoples in the south.

8. Frank M. LeBar, Gerald C. Hickey, and John K. Musgrave, eds., *Ethnic Groups of Mainland Southeast Asia* (New Haven, Conn.: Human Relations Area Files Press, 1964), 63 and 65.

9. Robert Ramsey, *The Languages of China* (Princeton: Princeton University Press, 1987), 278.

10. Ramsey explains the difference between the terms "Tai," "Dai," and "Thai" as follows: "*Tai* is the standard spelling for the name of the language family. It is also the collective term for the peoples who speak these languages. *Thai* and *Dai*, on the other hand, are the names of two specific Tai groups." Ibid., 232.

11. LeBar, Hickey, and Musgrave, *Ethnic Groups of Mainland Southeast Asia*, 65. For a recent ethnography of Guizhou's Miao, see Louisa Schein, *Minority Rules: The Miao and the Feminine in China's Cultural Politics* (Durham and London: Duke University Press, 2000). For a historical study, see F. M. Savina, *Histoire des Miao* (Hong Kong: Nazareth, 1930). See also Ruey Yih-fu, "A Study of the Miao People," in *Symposium on Historical, Archaeological, and Linguistic Studies on Southern China, South-East Asia, and the Hong Kong Region*, ed. F. S. Drake (Hong Kong: Hong Kong University Press, 1967), 49–58.

12. Lombard-Salmon, *Un exemple d'acculturation chinoise*, 152. Samuel Clarke mentions that depending on the location where they live they might also be called Yujia (around Anshun prefecture), or Shuijia (near Dushan), but that the term Tujia encompasses and includes Zhongjia, Yujia, and Shuijia. Samuel R. Clarke, *Among the Tribes of Southwest China* (London: Morgan and Scott, 1911; reprint, Taipei: Ch'eng-wen, 1970), 95.

as sinicized." [13] They have been equated with groups referred to as Shan in Burma, Lao in South China, and Tai in Thailand. [14] Today the former Zhongjia are called Buyi. They speak the Tai language.

According to the Qing Imperial Illustrations of Tributaries, "The Zhongjia Miao are the descendants of soldiers who guarded the frontier during the Five Dynasties period [907–960]." [15] Samuel Clarke, a missionary in Guizhou province at the turn of the twentieth century, also stated that the Zhongjia in Guizhou "invariably assert that their ancestors were Chinese who came from the province of [Guangxi], and many of them name the prefecture and county from which their forefathers came." [16] This claim to Chinese ancestry coincides with the view that they were descendants of Chinese troops sent to garrison the frontier areas. In any case their (ascribed) history as former defenders of the frontier may account for the interesting position they are often perceived to have had between the Chinese and the Miao.

Clarke articulated the following as one of several possible explanations for the name Zhongjia that he heard during his stay in China a century ago:

> The term [Zhongjia] is Chinese. *[Zhong]* possibly means the second of three brothers; *[jia]*, as we have already explained, means "Family" or "Tribe," and the term may be used to convey the idea that they are inferior to the Chinese and superior to the Miao. [17]

Whether or not this is the actual derivation of the term, the interpretation reinforces their status, in terms of acculturation, somewhere between the Han and the Miao. [18] Although there is some variation in orthography, the most frequent characters for Zhongjia are an easy shift from the characters Clarke discusses. [19]

---

13. Ramsey, *Languages of China*, 243.
14. Clarke, *Among the Tribes of Southwest China*, 89.
15. Chuang, *Xie Sui*, 585.
16. Clarke, *Among the Tribes of Southwest China*, 97.
17. Ibid., 95.
18. According to the *Ci Yuan* dictionary, Yuan Shu (?–199 A.D.), after occupying regions in the lower reaches of the Yangzi and Huai Rivers and declaring himself emperor in 197 A.D., took Zhongjia as his style name, or *hao*. Although Zhongjia as an ethnic category is not mentioned in the *Ci Yuan*, a possible connection between "Mr. Zhongjia" and the name of this tribe should not be completely ruled out. The tribe may be, or have been thought to have been, the descendants of his adherents.
19. The character *zhong* is often written with the dog radical. However, in *Qian ji* (1608) and the Kangxi Guizhou Gazetteer (1673), the person rather than the dog radical was used. Chinese mythology relates that south China's non-Han populations are the product of a union between a princess and a supernatural dog, Pan Hu.

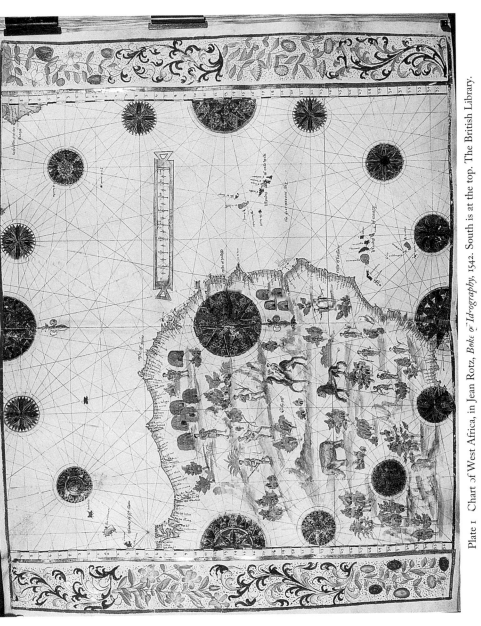

Plate 1 Chart of West Africa, in Jean Rotz, *Boke of Idrography*, 1542. South is at the top. The British Library.

Plate 2   Hei Luoluo. *Huang Qing zhigong tu* (Qing Imperial Illustrations of Tributaries). Detail.
National Palace Museum, Taipei, Taiwan, Republic of China.

Plate 3   Hua Miao. *Huang Qing zhigong tu* (Qing Imperial Illustrations of Tributaries). Detail.
National Palace Museum, Taipei, Taiwan, Republic of China.

Plate 4 Seventeenth-Century Dutch Cartouche, from De Wit's *Tabulae Maritimae ofte
ZeeKaarten,* 1675. Note "natives" in costume surrounding and facing the European who stands
taller than the rest. Courtesy The Newberry Library, Chicago.

Plate 5   Allegorical Frontispiece to Joan Blaeu's *Atlas Maior*, Amsterdam, 1662. Note what can only be the figure of Europa in the chariot. Figures representing other continents surround her on foot. Courtesy The Newberry Library, Chicago.

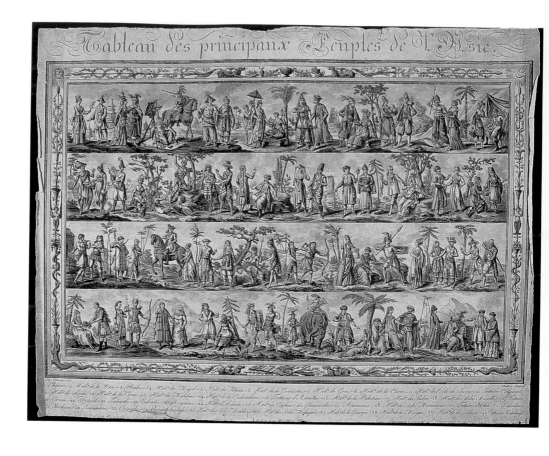

Plate 6   Les peuples de l'Asie. Jacques Grasset-Saint-Sauveur, *Tableaux des principales peuples de l'Europe, de l'Asie, de l'Afrique, de l'Amérique, et les découvertes des capitaines Cook, La Pérouse, etc. etc.* Courtesy The Newberry Library, Chicago.

The Dong Miao are portrayed here in a domestic setting. The illustration conveys information about the livelihood of the group and the skilled labor of which its members are capable. The illustration centers on a group of women around a loom, but a man and a child are also pictured. Note also the ornate seal script in which the text is written. The text reads as follows:

The Dong Miao are located in Tianzhu and Jinbing. They choose to dwell in level areas close to water, and raise cotton for a living. Many of the men hire themselves out to Han as laborers. Their clothing and eating habits are similar to those of Han. Women wear blue cloth head coverings in the shape of a horn and garments with patterned borders. They have the skill to weave fine brocade. Furthermore, the Dong Miao understand Chinese, and are docile.

Plate 8   Yao Miao. "Qian Miao tushuo" (album no. 57), entry 19. Courtesy of the Institute of History and Philology, Academia Sinica.

The illustration depicts a courtship ritual wherein young women have gathered in a raised bamboo building constructed especially for this occasion. Suitors, standing on the ground, serenade the women with reed pipe instruments. The text reads as follows:

Yao Miao are located in Pingyue prefecture. They are also called Yaojia. Many have the surname "Ji." Their disposition is yielding and agreeable. They are diligent and frugal. Although poor, they do not steal. Many attend school and sit for exams. Women are good at weaving and dying fabric. The first day of the second month of winter is their big festival. Furthermore, those who live in remote and inaccessible parts of the prefecture make upper garments by joining leaves together. They are as long as short skirts. When women reach the age at which they start to wear hairpins, a bamboo lodge is built for them in a remote place. Unmarried men come and play the sheng, *a reed instrument, to entice them. When someone dies the corpse is not buried. They use rattan to wrap the body, and put it in the forest.*

子吹笙誘之萵合死不羣以藤蔓束于樹間

短裙女子及笄造竹樓野處未婚男

俱笄處地方者樵木葉為衣着

為大節舟府屬之陳蒙爛土塞

讀書應試婦人善織染仲冬朔

姓性情柔順勤儉貧不為盗多

天苗在平越府又名天家多姬

Plate 9 Mulao. "Qian Miao tushuo" (album no. 57), entry 28. Courtesy of the Institute of History and Philology, Academia Sinica.

Two men and one woman appear outside their home engaged in religious activity. The woman beheads a chicken, presumably for sacrifice, and the men prepare a small altar. In front of the altar stands a straw dragon with five flags, each of a different color, in its back. The text reads as follows:

The Mulao have the surnames Wang, Li, Jin, and Wen. They are scattered in many different districts. In the winter they dig into the ground to make a stove. They make a fire in it and lie down around it. They do not use blankets or mats. They sacrifice to the spirits using a grass dragon into which they stick five colored flags. The sacrifice is performed in a nonresidential area. At festivals they sing and dance for enjoyment. Those in Duyun and Qingping dress like Han. When a close family member dies they wear coarse cloth, and do not use silk. The eldest son goes into heavy mourning; for seven times seven days he does not wash and does not go out of doors. If the eldest son is poor and cannot follow these observances, his eldest son or second son will substitute for him. They respect authorities and are strict in their instruction. Many of the young men enter school.

犵狫有王黎金文等姓散處各州縣
冬則掘地為爐唇火環卧不用被蓆
祀鬼用草龍揷五色旗往郊外祭之
遇節歌舞為歡在都与清平者衣
與漢同親死有斬衰而無苴絰居
居喪七七不沐浴不踰户若長子貧不
能守以長孫或次子代之尊敬師長
訓慕嚴子弟亦多入泮

Plate 10   Hong Miao (Red Miao). "Miao man tu" (album no. 40), entry 2. Courtesy of the Institute of History and Philology, Academia Sinica.

The illustration shows four men and four women in a natural setting. The women have placed themselves between the men in order to restrain them from fighting with one another. The Hong Miao are depicted in Miao albums as one of the fiercer Miao groups. The text reads as follows:

Hong Miao are located in Tongren prefecture. Many of them bear the five surnames Long, Wu, Shi, Ma, and Bai. They wear red silk clothing made by the women. The men tend to fight amongst themselves; the women must intervene to make them stop. On the third day of the fifth month married couples sleep separately. Furthermore they keep silent and stay indoors. These precautions are taken in order to avoid a kind of dangerous white tiger ghost. When they slaughter cattle they burn it to remove the hair. They boil the meat briefly, and also take the blood and drink it. When someone dies they use the clothes left behind by the deceased to make a likeness, and gather together to beat drums.

紅苗古銅仁布丈龍吳石麻白五姓
衣服悉用紅些女工以此為務若同顆
相閧必婦人解方罷五月寅日夫婦各
宿不敢言不出戶以避忌白帝也此殺
牲畜以火去毛微煮帶血食之人死將
故遺衣裝像衆皆擊鼓名曰吊故

Plate 11a   (Hei) Luoluo. "Bai Miao tu" (album no. 28), entry 1. Società Geografica Italiana.

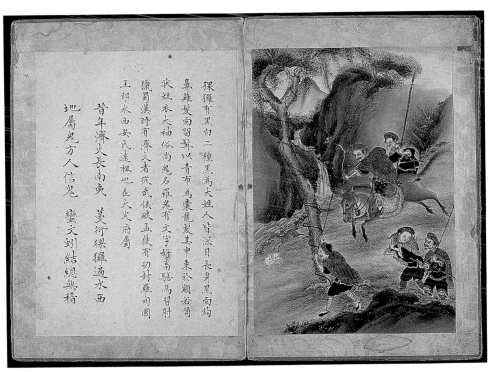

Plate 11b   (Hei) Luoluo. "Qian Miao tushuo" (album no. 56), entry 41. Harvard-Yenching Library, Harvard University.

Plate 12a Hua Miao.
"Bai Miao tu" (album no. 28),
entry 13. Società Geografica
Italiana.

花苗在大宝安順遵義貴陽阿屬皆姜姓氏其性强而
畏法俗酗酒不力勤女子用頭邑布相成八寶衣青布
衾男子吹蘆笙女子振鈴每歲孟春跳于平壊之地
日月場歌舞戲禮以終日暮則掣私以賤近共夕用
媒妁聘賢視女之姝為盈縮必男子女家掌婦成宿
同歸䘮明親族揚酒曲此賭環哭盡哀三七明携涯酒
俠遊宗延巫持兕謂之放又嵩畢磼𥔧𥔧礧僧之鬼散塟
不用棺槨不服棄惟宰牲禱鬼破家不悔版

Plate 12b Hua Miao. "Miao man tu" (album no. 40),
entry 24. Courtesy of the Institute of History and
Philology, Academia Sinica.

Plate 13a Bai Luoluo. "Bai Miao tu" (album no. 28), entry 3. Società Geografica Italiana.

Plate 13b Bai Luoluo. "Qian Miao tu" (album no. 53), entry 3. Courtesy of the Institute of History and Philology, Academia Sinica.

Plate 14 "Six Kinds." "Bai Miao tu" (album no. 28), entry 26. Società Geografica Italiana.

Plate 15a Jiugu Miao. "Bai Miao tu" (album no. 28), entry 40. Società Geografica Italiana.

九股苗在興隆衞凱里司與黑苗同類此種
固武侯南征戰之殆盡僅存九人遠嵩九股散
處蔓延地廣族繁性兇悍頗帶鐵盔前有護
面後無遮肩身披鐵甲至臍用鐵皮圍身鐵序
纏腿健者能左手執木牌右手持剝口嘲利刃
行走如飛撘箭藥弩名曰備架三人共張
弓無不貫雍正十年勒梅熏施搜繳兵器
建城設汛

Plate 15b Jiugu Miao. "Qian Miao tushuo" (album no. 57), entry 37. Courtesy of the Institute of History and Philology, Academia Sinica.

Plate 16   Jianfa Gelao. "Bai Miao tu" (album no. 28), entry 24. Società Geografica Italiana.

The second possible explanation Clarke relates is that *"[Zhong jia]* means heavy armor, and refers to the sort of armor used by them in ancient times."[20] This interpretation is less satisfactory because it relies exclusively on pronunciation, and not at all on the written characters.

Gelao

According to Lombard-Salmon, the appellation "Gelao" was first used during the Song dynasty, and names for subgroups within the Gelao category came into use during the Ming.[21] Today the Gelao number about 54,000. They are found primarily in northwestern Guizhou.[22] There is not complete agreement as to how to classify their language, but it may belong to a language group called Kadai, related to Tai.[23] Inez de Beauclair believes that the group may be related to the Lao (i.e. Man Liao) mentioned in records dating from as early as the Han dynasty. Tian Rucheng, a Ming dynasty author, indicates to his readers that the term Gelao was sometimes written as Geliao, *liao* being the same character used to indicate the ancient Lao mentioned in the Han texts of which Beauclair speaks.[24]

Lin Yüeh-hwa's "The Miao-man Peoples of Kweichow [Guizhou]" (a study that translates sections of the *Qiunnan zhifang jilüe,* prefaced 1847) states that the term Gelao derives from an earlier name by which this group, indigenous to what is now Guizhou and Guangdong, had been known in the Jin Dynasty (265-420 A.D.). At that time two groups fought each other. The victors assumed the appellation Zhuliao (Master Liao), and the defeated came to be referred to as the Pu, or enslaved Liao. Later the term Zhuliao evolved to become Gelao, and Puliao to become Mulao.[25] Jesuit accounts of early eighteenth-century Guizhou mention the Mulao, or "wood rats." They are identified as being better dressed than all the other Miao in the province. They may have been among those who worked as middlemen between the Sheng Miao and the Han in the timber trade.[26]

20. Clarke, *Among the Tribes of Southwest China,* 95–96.

21. Lombard-Salmon, *Un exemple d'acculturation chinoise,* 147.

22. Ramsey, *Languages of China,* 288.

23. LeBar, Hickey, and Musgrave, *Ethnic Groups of Mainland Southeast Asia,* 243. See also Ramsey, *Languages of China,* 288.

24. Lombard-Salmon, *Un exemple d'acculturation chinoise,* 148.

25. Lin Yüeh-hwa, "The Miao-man Peoples of Kweichow," *Harvard Journal of Asiatic Studies* 5 (1940): 275–276.

26. Du Halde, *Description . . . de la Chine . . . ,* vol. 1, 57.

Although little is known of the origins of the Gelao (who have retained this name under the People's Republic), it is clear that during the Ming and Qing the Han relied on them to form a kind of buffer zone against more threatening unpacified groups.

Luoluo

Luoluo was a derogatory term used to refer to a group now known as Yi who in the eighteenth century occupied a large domain including much of western Guizhou province. They had a highly developed system of self-government and complete control over both the routes through the area and the region's resource-rich mines. They were also famous for the quality of the horses they raised.[27] The Luoluo are still found in parts of Yunnan and Sichuan provinces as well as western Guizhou and northern Indo-China. Their language belongs to the Tibeto-Burman language family.[28] Although the term Luoluo has been replaced by Yi, I shall continue to use the former term here because this is how the group is identified in eighteenth-century texts.

The derivation of the term Luoluo, like Zhongjia, has numerous interpretations. According to the *Qiannan zhifang jilüe:*

> When Marquis Wu [Zhuge Liang] (181–234 A.D.) pacified the states of the South, he commanded all the heads of the great families to lead their own companies. [Luo Jihuo] of a great family of [Jianning], had his company in the region between [Zangge] and [Yelang]. This group was called by the name of [Luodian].
>
> At the close of the Sui (581–618 A.D.) and the beginning of the [Tang] dynasty, the capable leaders among the Man people were advanced to the status of [Guizhu]. The [Luodian] people then called the [Guizhu] of the [Luo] family by the abbreviated term of [Luogui]. This term was erroneously transformed into [Lulu] and later again into [Luoluo].[29]

According to this account the Luoluo were descended from a long and venerable line, although the more recent mutations of their name no longer reflected their origins. The change to Luoluo had apparently taken place already by the Yuan dynasty (1280–1386), although variations on writing the name continued throughout the Ming and well into the Qing dynasty.

Samuel Clarke provides us with a more popular late nineteenth-century account of the derivation of the name Luoluo:

27. Lombard-Salmon, *Un exemple d'acculturation chinoise,* 140–45.
28. LeBar, Hickey, and Musgrave, *Ethnic Groups of Mainland Southeast Asia,* 19.
29. Lin Yüeh-hwa, "The Miao-man Peoples of Kweichow," 272–74.

The term [Luoluo] is a name used for them by the Chinese, because they keep, or believe they keep, the souls of their parents in a miniature basket or hamper, just as the Chinese believe they have the souls of their ancestors in ancestral tablets. The Chinese word for such a basket is *[luoluo]*, and hence the name by which these people are generally known to foreigners. This miniature *[luoluo]* is about four inches deep and six inches in circumference, made of split bamboo, commonly wrapped round with a piece of calico, coarse as canvas, which is often the colour of a cocoanut from smoke and age.[30]

Soul baskets are not mentioned or illustrated in the Miao albums. Although the Bai Luoluo are often pictured with large oblong baskets strapped to their backs, there is no connection here with the small soul baskets.[31]

Clarke also mentions two other names for the Luoluo, Yijia and Yibian.[32] He further makes it clear that term Luoluo is offensive to the people it denotes, and that they prefer to call themselves Nosu.[33] Presently in the People's Republic of China they are officially called Yi,[34] although they still prefer to refer to themselves as Nosu.[35]

Yao

The appellation Yao has been in use since at least the Song dynasty.[36] In the early Qing many Yao lived in the western part of Hunan province, as well as in Guangxi. Many albums remark that Yao migrated into Guizhou province during the Yongzheng emperor's reign. The term Yaojia may have derived from a family name.[37] Today the appellation Yao covers a broad range of peoples who speak several different languages including Yao (or Myen), Punu (a kind of Miao), and Lakya

---

30. Clarke, *Among the Tribes of Southwest China*, 112–13.

31. For more on soul baskets, see Yang Fuquan, "The Ssú Life Gods and their Cults," in *Naxi and Moso Ethnography: Kin, Rites, Pictographs*, ed. Michael Oppitz and Elizabeth Hsu (Zürich: Völkerkundemuseum Zürich, 1998).

32. Because Clarke does not indicate the Chinese characters for these ethnonyms, they do not appear in the glossary.

33. S. Clarke, *Among the Tribes of Southwest China*, 113. For additional ethnonyms by which they have referred to themselves, see Victor H. Mair, "The Book of Good Deeds: A Scripture of the Ne People," in *Religions of China in Practice*, ed. Donald S. Lopez, Jr. (Princeton: Princeton University Press, 1996), 405. For more on the Yi, see Alain Y. Dessaint, *Minorities of Southwest China: An Introduction to the Yi (Lolo) and Related Peoples and an Annotated Bibliography* (New Haven: HRAF Press, 1980).

34. Lombard-Salmon, *Un exemple d'acculturation chinoise*, 139.

35. For an in-depth essay on the complex historiography of the Yi, see Stevan Harrell, "The History of the History of the Yi," in *Cultural Encounters on China's Ethnic Frontier*, ed. Stevan Harrell (Seattle and London: University of Washington Press, 1995), 63–91.

36. Lombard-Salmon, *Un exemple d'acculturation chinoise*, 157.

37. Lin Yüeh-hwa, "The Miao-man Peoples of Kweichow," 278.

(a Kam dialect belonging to the Tai family). Groups classified broadly by the government of the People's Republic as Yao have more than twenty different names by which they refer to themselves. The so-called Miao who live in Hainan are actually Yao according to linguistic classification.[38]

Zhuang

The Zhuang were a Tai group found in both Guangxi and Guizhou. Heavily sinicized, many of them had learned Chinese already in the eighteenth century. They were known for their cultivation of rice and use of water buffalo. The name Tu, or Turen, also referred to this group, perhaps because their livelihood was based on agriculture.[39] Today the Zhuang form the largest single non-Han ethnic group in China.[40] According to Eberhard, the Zhongjia, Longjia, Nong, and Dong all belong under the larger umbrella of Zhuang.[41] He also states that many of the people in south China who now identify themselves as Chinese stem from Zhuang groups but have now lost or deny their heritage.

Nong and Long

Nong are part of the larger Tai group, and can be categorized as a kind of Zhuang.[42] According to Luo Rao-dian the Nongjia zi were the descendants of the Nong family.[43] Long and Longjia, terms that appear in eighteenth-century works, do not appear separately in the reference works I have consulted on Chinese minorities. The terms may possibly

38. Ramsey, *Languages of China*, 286–87. For more on the Yao under the People's Republic of China, see the following, all by Ralph A. Litzinger: "Making Histories: Contending Conceptions of the Yao Past," in Harrell, *Cultural Encounters*, 117–39; "Memory Work: Reconstituting the Ethnic in Post-Mao China," *Cultural Anthropology* 13.2 (1998): 224–55; "Reimagining the State in Post-Mao China," in *Cultures of Insecurity: States, Communities, and the Production of Danger*, ed. Jutta Weldes, et al. (Minneapolis and London: University of Minnesota Press, 1999); and *Other Chinas: The Yao and the Politics of National Belonging* (Durham and London: Duke University Press, 2000). See also Jacques Lemoine and Chiao Chien, eds., *The Yao of South China: Recent International Studies* (Paris: Pangu, Editions de L'A.F.E.Y., 1991).

39. LeBar, Hickey, and Musgrave, *Ethnic Groups of Mainland Southeast Asia*, 230–31; Inez de Beauclair, *Tribal Cultures of Southwest China*, Asian Folklore and Social Life Monographs, vol. 2., ed. Lou Tsu-k'uang in collaboration with Wolfram Eberhard (Taipei: Orient Cultural Service, 1970), 70.

40. For a contemporary study of the Zhuang, see Katherine Palmer Kaup, *Creating the Zhuang: Ethnic Politics in China* (Boulder: Lynne Reiner Publishers, 2000).

41. Wolfram Eberhard, *China's Minorities: Yesterday and Today* (Belmont, Calif.: Wadsworth, 1982), 86.

42. LeBar, Hickey, and Musgrave, *Ethnic Groups of Mainland Southeast Asia*, 236.

43. Lin Yüeh-hwa, "The Miao-man Peoples of Kweichow," 278.

have been conflated with Nong due to the similar pronunciation of the two words. (Many native speakers of Chinese would not make or recognize the distinction between "n" and "l" that transliteration into *pin-yin* demands.)

Dong

Dongjia and Dongren were two names for the same group found in southeastern Guizhou and parts of neighboring provinces. The term Dong may have first appeared as recently as the Ming dynasty. The name Dongjia, according to some sources, derived from a family name.[44] Linguistically the Dong language is related to modern Tai dialects, but not very closely. It is actually part of a Kam-Sui dialect that diverged from Tai linguistic development.[45] Dong is the Chinese term for Tai peoples who refer to themselves as Kam.[46]

Jesuit Accounts

The Jesuit surveyors who traveled across all of China had opportunity to observe non-Han groups including both Luoluo and Miao. Their observations on the Miao give a contemporary, but outside, perspective on these non-Han groups:

> The Miao are spread out in the provinces of Sichuan, Guizhou, Huguang, Guangxi, and on the frontiers of the province of Guangdong. Diverse peoples are included under this general name, but the majority of them differ only in certain customs and relatively minor variations in language. The Miao of Sichuan, western Huguang, and northern Guizhou are like this. They are less gentle and less civilized than the Luoluo, and more at enmity with the Chinese.
>
> In order to make them submit, or at least to contain them, large garrisoned areas have been constructed at unbelievable expense. In this way success has been achieved in keeping them from communicating with each other. As a result the most powerful of these Miao are as if blocked off by forts and towns that cost the state a great deal, but which ensure the peace. . . .[47]
>
> The Miao who are in the center and southern part of Guizhou province differ from these. . . . Without getting hung up on the diverse names the Chinese have given them . . . one can divide them into Miao who have not submitted [to imperial authority] and Miao who have. This latter variety can be further subdivided into two kinds: those who obey Chinese magistrates and are in effect Chinese subjects, distinguishing themselves only by their

44. Ibid.
45. LeBar, Hickey, and Musgrave, *Ethnic Groups of Mainland Southeast Asia,* 231.
46. Ramsey, *Languages of China,* 233–34.
47. Du Halde, *Description . . . de la Chine . . . ,* vol. I, 55.

hairstyle . . . and those who have their own hereditary officials. . . . It was with the assistance of these Miao officials that the missionaries who worked on the maps of these provinces were able to learn something of the Miao of Guizhou who have not submitted. . . .[48]

All of the Miao are strongly disparaged in the minds of the Chinese. They say that they are inconstant, dishonest, barbarous, and above all notoriously thieving. This is something that Father Régis and the missionaries with whom he drew up the map of these provinces did not perceive. They found them, on the contrary, to be extremely trustworthy in returning the clothing which had been entrusted to them, attentive and self-applied to what they had been charged to do, hard-working, and willing to serve. But perhaps the Miao have their reasons for being discontent with the Chinese, who have taken from them almost all of the good land, and who continue to take possession of whatever they judge to be of benefit—if they are not made to stop by the fear of those whom they seek to despoil.

Whatever the case, it is certain that the Chinese neither like nor esteem the Miao and the Luoluo, and that these like the Chinese even less. They regard them as harsh and incommodious masters who keep them closed in by means of garrisons, enclaved as they are in the middle of them by a long wall that takes away all communication with other peoples from who they might draw assistance.[49]

These contemporary accounts by outsiders give a clear picture of the colonial nature of the "civilizing mission" and the tensions involved.[50] As such they provide good background for an examination of what the Chinese record says about the education and acculturation of Guizhou's non-Han inhabitants.

EDUCATION AND ACCULTURATION

Until the second half of the nineteenth century the majority of Guizhou's population was made up of non-Han peoples.[51] Over time various methods had been employed to incorporate these groups into the empire. One means was to promote assimilation through classical education based on Confucian values. Long before the Qing dynasty the Chinese defined their culture in terms of certain rites and other cultural practices vis-à-vis what they called "barbarian." Promoting the assimilation of these norms was seen as a crucial step in incorporating peripheral peoples into the empire and making them amenable to imperial

48. Ibid., 56–57.
49. Ibid., 60.
50. For additional Jesuit accounts of the Miao, see Savina, *Histoire des Miao,* especially chapter 2.
51. Lombard-Salmon, *Un exemple d'acculturation chinoise,* 170–71.

administration. This goal could best be achieved by providing an education based on Confucian values as expressed in the Confucian classics.

The Manchus adopted the legitimizing mantle of Confucian ideology as part of their strategy of rulership. However, a distinction needs to be made between the promotion of cultural assimilation to Han practices, which was pursued vigorously in the southwest during the reign of the Yongzheng emperor (1723–1735), and the use of Confucian ideology as a means to foster unity in a diverse empire as promoted by the Qianlong emperor. The former may have remained the goal of some Han officials even when official policy later shifted. As will be discussed in chapter 6, throughout the Qianlong reign the Miao albums constituted a celebration of (limited) cultural diversity under the umbrella of a protective empire where all people are fundamentally the same at the core, differing only in their outward physical manifestations.

Confucian standards of propriety and behavior were actively encouraged by Qing rulers within the interior (neidi). A document described as being "widely recognized as the most concise and authoritative statement of Confucian ideology"[52] was issued by the Kangxi emperor in 1670. Known as the Sacred Edict (shengyu), it was later reissued with an extensive explication in classical Chinese by the Yongzheng emperor in 1724.[53] The document consisted of sixteen maxims that codified and simplified the essence of what it was to be a good subject, so that these values could then be propagated both among rural, uneducated Han and among non-Han groups in territory newly incorporated into the empire. The basic precepts detailed proper behavior in relationships with others, thus laying the foundation for a smooth-functioning society. The Sacred Edict extolled the virtues of filial piety, brotherly submission, kindness, striving for peace, avoiding quarrels, cultivation of the mulberry tree (for raising silkworms), moderation, economy, respect for education, avoidance of heterodoxy, propagating the laws, behaving with propriety and courtesy, diligence in labor, instructing one's subordinates, not making or putting up with false accusations, not harboring wrongdoers, and paying taxes.[54] Both the Kangxi emperor's original text and the later explication promulgated by the Yongzheng emperor, as

52. Victor H. Mair, "Language and Ideology in the Written Popularizations of the *Sacred Edict*," in *Popular Culture in Late Imperial China*, ed. David Johnson, Andrew Nathan, and Evelyn S. Rawski (Berkeley: University of California Press, 1985), 325.

53. The Yongzheng reissue is known as the *Shengyu guangxun*.

54. See Mair, "Language and Ideology," 325–26, for a full translation of Kangxi's Sacred Edict.

well as a number of vernacular paraphrases of the Sacred Edict, were propagated widely among the common people, both Han and non-Han. The Sacred Edict also formed a central part of the curriculum in schools set up for non-Han children.[55]

During the first half of the eighteenth century, many officials advocated education as the preferred, albeit a long-term, means of promoting assimilation.[56] In 1705 the governor of Guizhou specially requested the establishment of public schools (yixue) for non-Han children. By 1741 no fewer than seventy-three schools had been established in the province.[57] Although it is not clear how many of these served minority populations, we do know that educating non-Han children was given special priority. In 1740 teachers in Guizhou's public schools received a salary of twenty silver taels per year paid by the central government;[58] schools in rural areas in the interior of China were, by contrast, funded by the local community and generally paid teachers in (copper) cash and grain.[59] The direct subsidy of schools in Guizhou shows a sensitivity on the part of the central government to the special need for education (and lack of local resources for it) in frontier areas.[60]

Zhang Guangsi, appointed governor of Guizhou province in 1729 and later made governor-general of Guizhou and Yunnan, was an eloquent advocate of promoting assimilation through education. His close association with the colonial project as a military figure, and later top administrator, gives his articulation of this goal special significance. In a Statement (to the throne) on Establishing Public Schools in Miao Territory, he set forth his views on Guizhou's ethnic groups and the

55. Ibid., 350–51.

56. China's examination system assured that officials were literati steeped in the Confucian classics. Theoretically, members of any ethnic group were eligible to qualify for the exams, but generally only those individuals backed with adequate family capital and therefore leisure time to study could pass—especially at the higher levels. Quotas were sometimes established for minority areas.

57. Lombard-Salmon, Un exemple d'acculturation chinoise, 261–62.

58. Evelyn S. Rawski, Education and Popular Literacy in Ch'ing China (Ann Arbor: University of Michigan Press, 1979), 57–58.

59. Ibid., 55.

60. An unusual album in the Musée Guimet, whose entries are arranged topically according to events rather than by ethnic subgroup, shows an illustration of Miao children attending such a school. A reproduction of this illustration appears in Lombard-Salmon's Un exemple d'acculturation chinoise, plate VIIb. For more on education in Qing China, see Elman and Woodside, Education and Society in Late Imperial China; Rowe's "Education and Empire in Southwest China," discusses the establishment of schools in minority areas in Yunnan. In Yunnan, unlike Guizhou, teachers' salaries were not funded directly by the central government. Rowe also notes that by 1751 some of Guizhou's schools in minority areas had been closed (443).

preferred means for their acculturation.[61] He affirmed the humanity of the Miao, stating that in both their physical constitution and mental awareness they were fundamentally no different from other subjects of the empire. Thus, he argued, it was possible to influence them through education (as Han too could be influenced).[62] His proposal also specifically mentioned the Yongzheng emperor's edition of the Sacred Edict as part of a suggested curriculum for Miao students.[63]

In 1729 the Yongzheng emperor followed up his reissue and explanation of the Sacred Edict with instructions for its dissemination. All towns and villages of any appreciable population density were required to establish locations that would serve to hold lectures on the Amplified Instructions.[64] Educating non-Han children in Confucian values presented special difficulties. As a result, sometimes instructors who could speak the native language were hired to teach in non-Han areas. Occasionally non-Han children attended schools along with local Han children where, administrators hoped, "gradually the culture of the latter would rub off on them."[65] Socializing the population into behaviors that produced good subjects was the ultimate goal. As Mair states, "[T]he education of non-Han subjects represented but an extreme form of the difficulties inherent in transmitting the values of the elite to the rest of the population." In this respect no distinction seems to have been made between Han and non-Han except perhaps in matter of degree.[66] In both contexts the Sacred Edict served to instruct commoners on how to be good members of an orderly and smooth-functioning society. In theory at least, the Confucian code of behavior propagated in the Sacred Edict had a universal appeal and application. The ideal was to civilize, not necessarily to sinicize.

THE TUSI SYSTEM

After 1644, when the Qing succeeded the Ming, it not only inherited the ideological models but also retained many of the administrative structures of the preceding dynasty. The *tusi* system, in which indigenous leaders *(tusi)* in frontier areas ruled over their own peoples, was

61. *Sheli miaojiang yixue shu,* GZTZ (1741), 60b–64a. A French translation appears in Lombard-Salmon, *Un exemple d'acculturation chinoise,* 293–95.

62. *GZTZ* (1741), 61b.

63. Ibid., 65a–b.

64. Mair, "Language and Ideology," 350. (As propagated by the Yongzheng emperor in 1724.)

65. Ibid., 350.

66. Ibid., 351. See also Rowe, "Education and Empire in Southwest China," 421.

one such institution. Begun as a convenient and relatively cost-effective way to bring additional territories and border populations under at least nominal imperial control, the system served as a basis for administering frontier areas populated by non-Han groups with minimal local disruption and military expenditure. Under the arrangement the state claimed the territory but did not have direct administrative responsibility for it.

The *tusi* were local headmen whose families historically had positions of authority in a given region, or individuals whose families had been rewarded for loyalty or service to the imperial government. *Tusi* were not all members of minority groups; sometimes Han Chinese were awarded these hereditary posts.[67] Unlike officials who were part of the regular Chinese system of administration, *tusi* could not simply be transferred to another region or another post. As a result, one drawback of the system for the imperial government was the difficulty involved in exercising authority over individuals with deeply entrenched local power.[68]

Upon its transition to power, the Qing government generally recognized those families who had held positions as *tusi* officials under the Ming, but not without careful consideration. To avoid conflicts over succession, genealogy had to be proven before new registry papers were issued.[69] In addition, heirs apparent "over thirteen had to take schooling in Han-Chinese ceremonial practices and were installed in office after they had passed tests."[70] Legitimacy had to be carefully established.

The process of incorporating new territory into the empire was a gradual one. Even after Guizhou was administratively recognized as a province in name, it was several centuries before it was brought under the direct administration of the central government—and even then such control was not complete.

GAITU GUILIU

During his reign, the Yongzheng emperor attempted to institute forceful imposition of central government control in Guizhou by replacing the *tusi* system with the regular system of administration used in the

67. Kent C. Smith, "Ch'ing Policy and the Development of Southwest China: Aspects of Ortai's Governor-Generalship, 1726–1731" (Ph.D. diss., Yale University, 1970), 301, note 1.
68. Lombard-Salmon, *Un exemple d'acculturation chinoise*, 212. For a close look at reforms to the *tusi* system under the Kangxi emperor, see John Herman, "Empire in the Southwest: Early Qing Reforms to the Native Chieftain System," *Journal of Asian Studies* 56.1 (1997): 47–74.
69. Wiens, *China's March toward the Tropics*, 226.
70. Ibid.

interior. The policy of changing from the *tusi* system and "returning" to a system of appointed rotating officials assigned by the central government is referred to succinctly in Chinese by the expression *gaitu guiliu*, literally "changing from the *tusi* system and returning to [the regular system of] rotating officials," or as Kent Smith translates it, simply "replacing native chieftains with regular officials."[71]

The debate over the merits of implementing the policy had started already in the Ming dynasty during the reign of the Hongwu emperor (1368–1399), when some officials advocated change, while others advocated retaining the *tusi* system. No major policy reversals were made at that time.[72] *Tusi* were replaced with Chinese officials only in cases where parties involved in serious disputes in *tusi* areas could not reach a settlement.

The debate continued unresolved into the Qing. Two separate, although sometimes overlapping, arguments favored bringing areas controlled by *tusi* under the regular system of administration. One claimed that the *tusi* were overly harsh and exploited those under their jurisdiction; because Chinese administration would be comparatively more just, its imposition would be both desirable and justified.[73] Furthermore, once the allegedly tyrannical *tusi* were suppressed, education could be introduced and the people made docile and good.[74]

The second, more overtly colonial argument cited the natural resources that could be more efficiently tapped if the areas in question were brought under the direct control of the central government.[75] Wang Lüjie, an official active during the Yongzheng period, outlined

71. Smith, "Ch'ing Policy and the Development of Southwest China," 42. The use of the term *gui*, to "return" or "go back," shows a Chinese predilection to invoke the past as a standard of the correct or proper order of things—even when there is no evidence that a past precedent existed for the ideal invoked.

72. For an in-depth analysis of the two competing views on whether or not to replace the *tusi* with the regular system of imperial administration, see Lombard-Salmon, *Un exemple d'acculturation chinoise*, 51–54.

73. According to Kent Smith, both the Yongzheng emperor and Ortai "clearly believed that in cases where native chieftains ruled their peoples oppressively, the imposition of direct Ch'ing administration would bring the local populations the benefits of peace and prosperity" ("Ch'ing Policy and the Development of Southwest China," 281).

74. Lombard-Salmon, *Un exemple d'acculturation chinoise*, 213.

75. Guizhou's known natural resources in the eighteenth century consisted primarily of forests and mineral reserves. For specifics on species of trees native to Guizhou, see ibid., 86–89. Lead ore, cinnabar, and iron ore were found throughout most of Guizhou (ibid., 91). In addition copper, silver, and orpiment (sulphur of arsenic) were found in various regions. Lombard-Salmon concludes that these rich mineral resources made Guizhou among the most resource-abundant provinces in of all of China (91–92).

five reasons in favor of *gaitu guiliu*. Briefly, they revolved around strategic considerations, forest resources, metals (from mining) and their potential profit for the treasury, the civilizing mission, and the possibility of assimilation of the Miao over time.[76] This argument focused around the economic benefit of colonizing the frontier. Proponents of *gaitu guiliu* also focused on the attractiveness of *tusi* areas as a refuge for Chinese lawbreakers.[77]

In spite of Guizhou's inhospitable conditions, by the eighteenth century at least two factors made the arguments for bringing Guizhou under the control of the central government more compelling. These were a dramatic population increase, due largely to an influx of Han settlers, and the province's strategic position. It was necessary to cross Guizhou in order to reach resource-rich Yunnan province.

Although accurate population records do not exist for Guizhou in the eighteenth century due to drastic under-reporting,[78] the effects of population growth and resulting migration doubtless affected the province at this time. A burgeoning mining industry drew many people to the southwest,[79] and famine chased others there from their home regions. Tradespeople from other parts of China also set up shop in Guizhou. The variety of local origins represented among Guizhou's growing population during the eighteenth century can be seen in the case of Weining, where there were immigrants from Jiangnan, Huguang, Jiangxi, Fujian, Shanxi, Yunnan, and Sichuan provinces.[80] The introduction of new-world crops that could grow in poorer soils, such as sweet potatoes, peanuts, and corn, helped to support Guizhou's growing population.

Guizhou's population growth notwithstanding, the most pressing reason for incorporating the territory now known as Guizhou into the empire was to allow for the passage to Yunnan necessary to ensure that province's continued submission to imperial control. Access to Yunnan became especially crucial from the 1720s. Up until 1715 the Qing government had procured copper from Japan, but after this time Japan no

---

76. Ibid., 216–217.

77. Smith, "Ch'ing Policy and the Development of Southwest China," 251, and 44–45; and Lombard-Salmon, *Un exemple d'acculturation chinoise,* 213.

78. According to Ho Ping-ti's *Studies on the Population of China,* non-Han populations under *tusi* control in the Southwest were not recorded prior to, or during, the eighteenth century (9, 50–51, 287). In addition, in many areas of Yunnan and Guizhou where the population was a mix of Han and non-Han peoples, figures also tended to be under-reported, if reported at all (50–51).

79. Susan Naquin and Evelyn S. Rawski, *Chinese Society in the Eighteenth Century* (New Haven and London: Yale University Press, 1987), 200.

80. Lombard-Salmon, *Un exemple d'acculturation chinoise,* 170–71.

longer met the demand, and other sources had to be found. Shortly thereafter rich reserves were discovered in Yunnan province, and by 1738 the mines there supplied the full coinage needs of the Chinese government.[81]

In this context Ortai, governor-general of Yunnan and Guizhou from 1726 to 1731 and a close confidant of the emperor, helped to convince the Yongzheng emperor to move ahead with the new policy of *gaitu guiliu*. He had first memorialized the throne on the topic of the annexation of "Miao territory"[82] in the spring of 1726. By the beginning of 1728 he had made extensive plans for its annexation. The emperor, persuaded by the case Ortai made, gave his assent. The result was a series of military campaigns that lasted several years, from 1728 to 1731. Although briefly successful, the Yongzheng campaigns were followed by widespread revolt.

POLICIES OF ASSIMILATION UNDER ORTAI

Ortai focused his efforts on Miao territory in southeast Guizhou, with varying degrees of force proving necessary to meet his goals. In the following pages I concentrate not on his military advances, but on the combination of additional measures, both actively implemented and passively permitted, that promoted the more thorough incorporation of Guizhou into the empire. Han settlement, forceful attempts at cultural assimilation, and institutionalized discrimination against the non-Han indigenous population all contributed to Guizhou's colonization.

The encroachment of Han people onto lands formerly occupied by other ethnic groups was not new, but this type of conflict intensified around the turn of the eighteenth century when population pressures aggravated already existing tensions. The Yongzheng emperor's revival of a system known as *tuntian* exacerbated the problem. Under this policy lands were forcibly wrested from non-Han peoples who had cultivated them. Soldiers took over the production of these lands thereby helping to earn their own keep while simultaneously maintaining terri-

---

81. Ibid., 165.

82. In the context of early eighteenth-century Guizhou, the term *Miao jiang* (Miao territory) referred to an area of southeastern Guizhou inhabited almost exclusively by non-Han peoples. Until Ortai's attempts to annex the area, central government administration had not penetrated into the region (Smith, "Ch'ing Policy and the Development of Southwest China," 249). The term can also be used generally to speak of areas populated primarily by Miao.

torial gains. Parts of southeast Guizhou near the border with Guangxi and Hunan, including the rivers that linked these provinces, were taken in this way. We know that at least some officials felt that there was no policy problem with turning such lands over to Han civilian cultivation after they had been secured for some time by the army.[83]

Han settlers also found additional and more creative ways to acquire land. Perhaps most commonly, lands were lost to Han creditors when debtors were unable to meet their obligations. Some officials recognized this appropriation of land as a source of local discontent, and eventually measures were taken to limit this problem.[84] In other cases Han men married into local non-Han families and thus inherited land. In 1727, one official voiced the opinion that such marriages should not be allowed. In a memorial to the court, he explained that they only caused difficulties in the long run by creating situations in which the Han would "establish hideouts among the Miao, and the Miao would use the Han as spies." Additionally he disapproved of the children of such marriages "modeling themselves entirely on the habits of the family of their mother, considering such things as abductions and assassinations as everyday affairs."[85] It should be noted that later (in 1764), intermarriage was, by contrast, advocated as a means to promote assimilation of aboriginal groups to Han culture.

Institutionalized discrimination also played a part in the colonization of Guizhou. One regulation put forward c. 1703 stated the following: If a Miao should kill a Han, then the lives of two Miao would be taken in compensation. If a Han were raped by a Miao, the whole family would be seized in compensation. If the Miao continued to engage in subversive activities rather than devoting themselves to agriculture, then the army would be sent to wipe them out. If a quarrel resulted in murder, no investigation would be carried out, rather Qing troops would attack the village where it happened. Threats and force were the order of the day. Later, under Ortai, it became standard procedure to confiscate all weapons and prohibit the manufacture of new ones. Any armed persons were considered "Miao rebels" and were liable to be seized and executed.[86] Under the Yongzheng emperor the Miao were also told that,

83. Lombard-Salmon, *Un exemple d'acculturation chinoise,* 173.

84. Ibid., 174–75. John Shepherd also addresses this phenomenon in Taiwan. See his *Statecraft and Political Economy on the Taiwan Frontier, 1600–1800* (Stanford: Stanford University Press, 1993), especially chapter 9, "The Evolution of Aborigine Land Rights."

85. Lombard-Salmon, *Un exemple d'acculturation chinoise,* 225.

86. Ibid., 221.

once census records were established, they would have to pay tax and participate in the corvée labor system. If they did so they would be treated as Han, if they refused they would be considered as Miao rebels and suffer the consequences.[87] This policy of taxation was reversed under the Qianlong emperor.

Harsh laws and threats were not the only indications of Guizhou's status as a kind of colonial space in need of special measures. The simultaneous need to protect indigenous populations from official exploitation also points to the fact that Guizhou—as a multiethnic frontier region in the process of being made into part of the regular system of administration—still required special consideration with regard to administrative policy. For example, the Yongzheng emperor decreed in 1727 "that, in areas newly changed from native to regular administration, officials who sent yamen runners and soldiers to harass the people should be punished more heavily for the crime than officials in the interior."[88] In the following year he once again warned that officials should not try to exploit the natives:

> I fear the attitude of the officials is that because these natives were formerly under oppressive rule [of the *tusi*], now that they have come over to us and cast off their former tyranny, there is no harm in extorting a bit from them. This point of view is absolutely impermissible.[89]

These decrees provide evidence that the state recognized the precariousness of imperial domination in these newly colonized areas.

Other special administrative arrangements also call attention to Guizhou's status as part of China's newly incorporated internal frontier. Responding to a request from Ortai, the Yongzheng emperor agreed that each assistant magistrate assigned to a post in the newly established subprefectures have personal command over one hundred soldiers.[90] This request, and its fulfillment, reveals a certain level of wariness, if not fear, among Guizhou's officials about the need for protection from a potentially hostile population. Taken together, the examples of administrative regulations reveal Guizhou's position as a newly colonized area of China.

Although, as discussed above, moderates advocated relatively "peaceful" means of acculturation through education and the propagation of

---

87. Ibid., 227–28.
88. As quoted in Smith, "Ch'ing Policy and the Development of Southwest China," 285.
89. Ibid., 284.
90. Ibid., 273.

the Sacred Edict in minority areas, the achievement of a certain degree of cultural conformity in non-Han areas was also valued by the "hardliners," and enforced brutally. Ortai, after encountering repeated resistance, decided harsher measures were necessary. He not only confiscated the rebels' arms but also "built new military posts, occupied strategic points, [and] obliged the indigenous populations to cut their hair and change their (style of) clothing."[91] At least in certain areas, male Miao were not allowed to wear traditional garb, but were forced instead to dress like their Chinese neighbors. This included their hair, which was to be worn in a queue with the forehead shaved, rather than in a topknot.[92]

In addition, both indigenous beliefs and festivals were attacked as vulgar and worthy of suppression. One of the reasons given for the suppression of certain rites was their onerous expense.[93] Another more fundamental reason was that by forbidding such festivals the state took away both a reason and an opportunity for the Miao to come together in large numbers, which would naturally have made officials nervous. The prohibition also denied non-Han peoples of an important means of reaffirming their cultural identity.[94]

The indigenous residents of the areas targeted by Ortai's campaign exhibited determined resistance to the military advances directed toward them.[95] By 1731, however, Ortai had met his territorial goals, and walled towns had been built to house seven new garrisons in southeastern Guizhou.[96] By all appearances the governor-general had successfully completed his campaign, if with more effort and bloodshed on both sides than he had initially envisioned. However, perhaps due to in part to the harshness with which they were carried out, Ortai's string of victories soon proved to be short-lived. In 1735 the Miao in the newly conquered region rebelled fiercely. The rebellion even spread to areas that had been under imperial control prior to Ortai's campaigns.[97]

91. Lombard-Salmon, *Un exemple d'acculturation chinoise*, 232.

92. Ibid., 222. The queue was imposed on the Han by the Manchus after the conquest. See Philip A. Kuhn, *Soulstealers: The Chinese Sorcery Scare of 1768* (Cambridge, Mass.: Harvard University Press, 1990), for the importance of the queue as a political symbol of submission to the Manchu government.

93. *Miao fang bei lan* as quoted in Lombard-Salmon, *Un exemple d'acculturation chinoise*, 221; see also 222.

94. Ibid., 222.

95. Smith, "Ch'ing Policy and the Development of Southwest China," 263.

96. Ibid., 273.

97. Ibid., 275–76. These areas were vulnerable because during the pacification campaigns forces had been shifted from Chinese towns to areas deep within Miao territory, leaving the former places

The Yongzheng emperor died in 1735 leaving the Qianlong emperor to deal with the rebellion. Prior to his assumption of power at age twenty-five, the Qianlong emperor had been a member of the "Council of Princes and Ministers to Manage Miao Territory Affairs" created by the Yongzheng emperor in 1735. Therefore he was already familiar with the situation in southwest China.[98] He ordered Zhang Guangsi to retake the area, a mission accomplished in 1736.[99]

On the whole, however, the Qianlong emperor tended to eschew direct military confrontation with the Miao as an ongoing policy, preferring comparatively less aggressive means of dealing with them. Once the area was reconquered, the new emperor made some changes in its administration. Taxes would not be collected in the newly annexed territory, and Miao disputes would be settled not according to Chinese legal methods, but by Miao custom.[100] This method of handling the rebellion shows equally the desire on the part of the new emperor for preserving the appearance of Miao submission to imperial control, and a willingness to achieve a viable solution through compromise. The rebellion that sprang up in the wake of the Yongzheng suppressions had shown that brute force alone was not an effective, or at least cost-effective, means to integrate this area into the empire. Other less confrontational means now appeared preferable. The Qianlong emperor's reign saw the advancement of other means for governing more effectively, namely the accumulation of information about the various peoples populating the empire. Ethnographic portions of local gazetteers and the Miao albums are one manifestation of this practice. The information they contained would assist officials assigned to governing frontier areas.

---

virtually undefended. Thus at the height of the rebellion the central government was in control of less territory than prior to Ortai's campaign.

98. Harold L. Kahn, *Monarchy in the Emperor's Eyes: Image and Reality in the Ch'ien-lung Reign*, East Asian Series, no. 59 (Cambridge, Mass.: Harvard University Press, 1971), 110.

99. Smith, "Ch'ing Policy and the Development of Southwest China," 278–79. Victory, however, is relative. For example, although regular markets were established in the locality of Guzhou at this time, it was not until 1786 that Chinese would dare to enter parts of inner Guzhou to cut wood. Beauclair, *Tribal Cultures of Southwest China*, 68–69.

100. Smith, "Ch'ing Policy and the Development of Southwest China," 279.

CHAPTER FIVE

The Development of Ethnographic Writing in Guizhou Province, 1560–1834

"If one does not differentiate between their varieties, or know their customs, then one has not what it takes to appreciate their circumstances, and to govern them."
—preface to "Man liao tushuo"

his chapter traces the emergence of ethnographic inquiry in Guizhou province during the early modern period and, more specifically, the relationship between ethnographic depiction and colonization in what has become southwest China. Ethnographic writing found in local gazetteers and histories of Guizhou from 1560–1834 shows a correlation between imperial expansion and the development of the ethnographic record. In the transition from generally descriptive to more strictly ethnographic accounts of non-Han peoples, we see an epistemology that valued measurement, precision, and "objective" description taking shape.

We can trace four significant ways in which classical Chinese ethnographic texts became more comprehensive and the system of classification around which they are organized more intricate up until the mid–eighteenth century. Perhaps most important for the argument that ethnographic depiction and colonization were closely related is the simple fact that in more recent editions of the Guizhou provincial gazetteer the territorial scope covered in the ethnographic descriptions of Guizhou's people expands, revealing increased bureaucratic penetration into frontier regions. Not surprisingly, when new groups are mentioned for the first time in a revised gazetteer, the administrative district in which they are said to dwell is also mentioned for the first time. In other words, as the Qing was able to extend its administrative reach into more remote areas, the scope of geographical inquiry reflected in the provincial gazetteers increased. Secondly, we also have evidence that, as territorial reconnaissance and control improved, compilers increasingly included information based on direct observation rather than relying only on previous textual sources. Gradually, additional detail was added to existing entries as more information became available. Related to the first two trends is an increasing refinement and systematization in the categorization and naming of non-Han groups residing in Guizhou seen in later editions of the provincial gazetteer. The number of different groups that was recorded by name, and thus categorized, increased steadily. The number grew from thirteen in 1608 to thirty in 1673 to forty-one in 1741 to eighty-two by 1834.[1] Finally, administrative concerns, although present throughout, are gradually represented as goals achieved.

SOURCES

Until recently, change over time in Chinese ethnographic texts from the Ming and Qing periods had not been seriously studied. Several scholars used these documents as sources for ethnographic or historical information, but they left unasked the question of the development of ethnographic writing.[2] Close reading shows that the criteria for ethnographic writing about non-Han groups in Guizhou underwent significant revision, gradually becoming more "ethnographic," or scientifically based, during the course of approximately three centuries. The unfolding of a similar process has been identified in Guangxi province during the late Ming.[3] These findings parallel those of Hodgen for the early modern European context.

My observations are based on the analysis of the relevant chapters (*juan*) of six works on Guizhou chosen for the extensive descriptions

1. Although the earlier, 1560, text discussed sixteen different groups, they did not all reside in Guizhou.

2. See, for example, Wolfram Eberhard, "Die Miaotse Alben des Leipziger Völkermuseums," *Archiv für Anthropologie* 26 (1941): 125–37; Beauclair, *Tribal Cultures of Southwest China;* and Lombard-Salmon, *Un exemple d'acculturation chinoise.* Lombard-Salmon does discuss advances in knowledge of geography and mapping as found in gazetteers, but not in relation to ethnographic knowledge. See especially 69, 245–46, and plates 1–3.

3. Leo Kwok-Yueh Shin, "Tribalizing the Frontier: Barbarians, Settlers, and the State in Ming South China" (Ph.D. diss., Princeton University, 1999).

they contain of Guizhou's non-Han groups. Each chapter is arranged according to entries headed by the names of the various non-Han groups resident in the province. A significant degree of intertextuality exists; more recent accounts used the older accounts as source material. My focus is on the changes that constitute the emergence of ethnographic writing over time.

The texts are: Tian Rucheng's *Yanjiao jiwen* (The Southern Frontier: A Record of Things Heard), 1560; Guo Zizhang's *Qian ji* (Record of Guizhou Province), 1608; three editions of the *Guizhou tongzhi* (Guizhou Gazetteer), 1673, 1692, and 1741; and Li Zongfang's *Qian ji* (Record of Guizhou Province), prefaced 1834. Of the six works, the three gazetteers were officially commissioned; the local histories were written by private individuals who also happened to hold government office.[4]

Yanjiao jiwen (1560) was written by Tian Rucheng during the Ming Dynasty (1368–1644). The author received his *jinshi* ("Presented Scholar" degree), the highest-level degree in Ming China, by passing the highly competitive national civil service exams in 1526. As a successful candidate he qualified for government office. Tian served in a variety of posts, some of which took him to parts of southwest China, including Guizhou. During his tenure there he developed interests in local history and the native peoples that are reflected in his writing. The latter pursuit was not purely academic; in the course of his career he had been involved in suppressing tribal rebellions.[5] In his *Yanjiao jiwen* he expresses his opinions on the proper handling of non-Han groups.

Qian ji also dates from the Ming, but almost fifty years later (1608). This local history was compiled by Guo Zizhang, a native of Taihe in Jiangxi province. Guo Zizhang obtained the *jinshi* in 1571. His official posts included Provincial Director of Education in Sichuan province (1586–1589), and Governor of Guizhou (1599–1608). As governor he was involved in quelling a Miao rebellion in Bozhou.[6]

Guo's scholarly pursuits included the study of geography. As discussed

4. I use the term "local histories" for books written by individuals about a given region *(ji, shu)*. I reserve the term "gazetteer" *(fangzhi,* or *tongzhi)* for works about a given region of a more official nature; they were often imperially commissioned. The distinction is in some ways minor as the same individuals were often involved in the compilation of both. However, the gazetteers tend to follow a more standardized format in terms of the content included.

5. L. Carrington Goodrich, ed., *Dictionary of Ming Biography, 1368–1644,* the Ming Biographical History Project of the Association for Asian Studies (New York and London: Columbia University Press, 1976), 1286.

6. Bozhou has been called Zunyi since 1728 when it was newly incorporated into Guizhou from Sichuan province. It is located in the north-central part of Guizhou.

in chapter 2, Guo had come to know Matteo Ricci, an early Jesuit missionary to China, while living in Guangdong province. They discovered a shared interest in geography; Guo later reprinted Ricci's world map in one of his own works.[7]

From the mid–seventeenth to mid–eighteenth centuries, imperially commissioned, officially compiled gazetteers of Guizhou province provide the richest source of information about its non-Han groups. Editions date from 1673, 1692, and 1741. The earliest edition was imperially commissioned by the Kangxi emperor in 1672.[8] The ethnographic texts that concern us are found in chapter 29, *tusi*,[9] under the subheading *miao liao*, or "southern barbarians." Here, for the first time in a gazetteer of Guizhou, woodblock illustrations appear next to the textual entries describing the various non-Han groups that populated the province.

The 1692 version of the Guizhou Gazetteer, as detailed below, revises and adds to the information contained in the earlier edition. The most striking difference is the revised sequence in which the groups are mentioned, and major changes in the woodblock illustrations that accompany the texts. Not only does this revision show a continued interest in working with the ethnographic information, it reveals a direct connection with the emergence of the Miao album genre discussed in detail in chapter 6. The new sequence in which groups are mentioned parallels that found most often in the later albums. Furthermore, not only do the new illustrations show a refinement in quality from the 1673 edition, but their actual content and composition closely parallels that of the Miao albums. The activities in which specific groups are engaged, the variety of positions the figures assume, and often even their placement on the page, closely resemble illustrations in many of the albums.

Almost fifty years later another revised version of the Guizhou gazetteer appeared. The work was commissioned by the Yongzheng emperor in 1729, when Ortai was governor-general of Guizhou and Yunnan provinces, but completed only in 1741, when it was submitted by a succeeding governor-general, Zhang Guangsi.[10] Unlike the Guizhou gazetteers compiled during the Kangxi emperor's reign, this edition does not include woodblock illustrations of Guizhou's native peoples. The

---

7. Goodrich, *Dictionary of Ming Biography*, 775–77. See also D'Elia's detailed account of this relationship in "Il mappamondo Ricciano Nell'atlante de Cuozzimluo (Kweiyang, Kweichow, 1604)," in D'Elia, *Il mappamondo cinese*, 73–81.

8. Jaeger, "Über chinesische Miaotse-Albums" (1917), 88.

9. See chapter 4 for a discussion of the *tusi* system.

10. Jaeger, "Über chinesische Miaotse-Albums" (1917), 85–86.

change may reflect a decision not to compete with the superior hand-painted color illustrations of the emergent Miao album genre. The texts are similar in general style and content to the earlier editions, but they are more extensive, and there are more of them.

Nearly a century intervened between the 1741 Guizhou Gazetteer and Li Zongfang's *Qian ji*, prefaced in 1834. Li tells his reader that the third chapter of *Qian ji* was based on a Miao album containing eighty-two entries that was made by a former official posted to Bazhai, a small transport station located in Duyun prefecture. Following the last entry a note states:

> The "Illustrations of eighty-two varieties of Miao" *(bashier zhong miao tu)* was made by Chen Hao, Bazhai's former Assistant Prefect in charge of the Miao.[11] I had heard that the printing blocks were kept in the frontier office. But at present they are no longer (there).[12]

Li continues to explain that *Qian shu* (1690) (a local history whose ethnographic content is nearly identical to the 1673 Guizhou Gazetteer) contained entries for thirty varieties,[13] but that Chen Hao's illustrations and explanations were more detailed and exhibited more care than those of the earlier work. However, Li adds that he has chosen to additionally polish and refine Chen's text. He notes that the later texts record all of the groups named by Tian Wen in *Qian shu*, but do not repeat all of the information contained in each entry. He does not discuss the basis for his changes aside from the "polishing and refining" of the language.

Across the six texts we see a substantial amount of borrowing. However, up until at least 1741 the content is not mindlessly repeated, but carefully sifted through, revised, and consistently expanded. The 1834 text, by contrast, shows a reverse in this trend. Although the number of categories of non-Han peoples has doubled, reaching eighty-two, the amount of text in each entry is severely reduced. By the mid-nineteenth century the careful search for direct information based on observation seems no longer to have been pursued. Causes for this reversal may stem from the general political situation at the time. The Western powers were beginning to reveal themselves as a serious threat, Opium imports from British India were growing at a rapid pace, and internal rebellion

---

11. *Li miao tongzhi.* See H. S. Brunnert and V. V. Hagelstrom, *Present Day Political Organization of China,* trans. A. Beltchenko and E. E. Moran (Shanghai: Kelly and Walsh, 1912), entry no. 849, pages 427–29.

12. Li Zongfang, *Qian ji* (1834), 3:9b.

13. Because the ethnographic portion of the 1690 text is so close to the 1673 gazetteer I do not discuss it further here.

threatened the Qing from within by the early nineteenth century. Energy and resources were needed elsewhere.

GEOGRAPHICAL EXPANSE

Tian Rucheng's *Yanjiao jiwen* deals not only with Guizhou, but with the southern frontier more generally as indicated in the title. The author shares his vision of what the future might hold for the region about which he writes, revealing much about the reasons for his interest:

> How do we know that after 100 generations Yunnan's territory will not have a glorious reputation and human knowledge like that which was transmitted to Fujian and Guangdong? How do we know that the barbarians *(yi)* in Miandian, 800 li away, will not have a day when prefectures and counties are enumerated and officials are sent [to rule them]?[14]

This passage foreshadows to a large degree the events that unfolded during the subsequent development of the area. Although the Chinese did not colonize Burma, in the coming centuries the once remote regions of Guizhou and Yunnan were increasingly controlled by central authority. Later, gazetteers, Miao albums, and the Qing Imperial Illustrations of Tributaries made of the region did enumerate these various regions and counties as well as a great deal of information about their inhabitants, who were being incorporated more tightly into a unified but diverse empire. *Yanjiao jiwen* shows that as early as 1560 there was an interest in the inhabitants and aspirations for territorial control of this region.

A comparison of Tian Rucheng's *Yanjiao jiwen* (1560) with *Qian ji* (1608) already reflects an increase in the territorial scope of ethnographic inquiry within Guizhou province. By 1608 entries had been expanded to include additional place names.[15] The geographical scope of *Qian ji*'s investigation is also more specific in its references to geography and includes more area within Guizhou. For those groups described in both the 1560 and 1608 texts, no fewer than twenty-nine place names appear for the first time in *Qian ji* (1608), representing a significant geographical expansion in knowledge regarding the non-Han groups of

14. Tian Rucheng, *Yanjiao jiwen,* Yingyin wenyuange siku quanshu edition, vol. 352, p. 658. Miandian is roughly equivalent to Myanmar (Burma) today.

15. The 1608 Yanghuang entry, in addition to mentioning all the place names mentioned in Tian Rucheng, also lists as many as eight others: Duyun, Liping, Tanqi, Xinhua, Zhonglin, Liangzhai, and Hu'er. The 1608 (Si) Longjia entry includes the place name Dingying, as well as retaining Ninggu and Xibao.

Guizhou.[16] In the southwest part of the province information is mentioned for the first time about a group (the Zhongjia) as far away as Pu'an, a point directly along the major east-west artery leading into Yunnan. To the southeast, in Liping prefecture, as many as twelve new place names are mentioned; the 1560 text named only one, Tanqi township. In central Guizhou too, many more places are mentioned, including Guiyang, Xintian, Liping, Longli, Pingfasi, Bangshui, Duyun, Dingfan, Changshun, and Danxing. Finally in the east central region Xinglong, Bianqiao, Zhenyuan, Angshui, and Miaomin are mentioned for the first time. The mention of these additional place names demonstrates a substantial increase in the areas of Guizhou that were accessible to those interested in learning about its native peoples over the course of the late Ming.[17]

Expansion of geographical knowledge accounts for the inclusion of additional groups in 1608. New groups appearing concurrently with new place names are the Turen, Manren, Dongren, and Yangbao (see table 1). Many of the place names mentioned under the new entries appear for the first time in 1608—Zunyi (Yangbao); Angshui, Caodidong township in Liping prefecture (Turen); Dingying in Anshun prefecture (Silongjia), Danxing in Duyun prefecture (Manren); Yongcong military post *(zhai)* and Hongzhou (Dongren). The most likely reason that these groups and places were not mentioned in the earlier text was because they were previously inaccessible to members of officialdom (and perhaps to Han Chinese in general).

A trend toward geographic expansion in ethnographic inquiry continues in the 1673 Guizhou Gazetteer, which contains a total of thirty entries. The seventeen new entries contain a total of four place names mentioned for the first time: Shibing, Guiding, Guangshun, and Zhenning. These four places are all relatively distant from each other and span the province east to southwest. They do not represent increased connaissance in one geographical region only, but rather reflect a larger,

---

16. Most entries contain more than one place name.

17. Of those groups with entries in both texts there are only two place names mentioned in 1560 that do not reappear in 1608. They are Yanhe youqi and Wuchuan, both located in the extreme northeast of the province (Sinan prefecture). These places appeared in *Yanjiao jiwen* under the Ranjia entry. The text says that the Ranjia lived between these two places. The later, 1608, text indicates that the Ranjia lived in Yanhe, which is located very near to the places mentioned in 1560 between which the Ranjia were said to live. The 1608 Ranjia entry also mentions a second location, Shiqian, located directly south of Yanhe, in Shiqian prefecture. Thus, although some specific place names do not recur in the later text, the geographical expanse covered by the place names mentioned is matched, and indeed greatly expanded, in *Qian ji* (1608).

more widespread effort at increased information gathering. Three of these four places, Shibing, Guiding, and Zhenning, are located directly on the main east-west road leading through Guizhou. Only Guangshun is located off of this road to the south. Again, an increase in newly mentioned territory corresponds with information about additional groups. Furthermore these areas were all in strategic locations.

New place names also appear in entries for groups that were previously mentioned. We see a marked expansion of recorded place names in the western portion of the province. All the place names appearing in Dading prefecture are new in 1673, reflecting administrative changes; Dading prefecture was established 1665 when the boundary of Guizhou province was moved westward to incorporate some of what had earlier been Sichuan province.[18] Two new place names in the northeast also appear: Tongren in Tongren prefecture, and Sizhou in Sizhou prefecture. In the south the scope of ethnographic inquiry is expanded to Danping in southern Guiyang prefecture.[19] The gazetteer also more consistently records the administrative status of place names mentioned, i.e., county, department, or prefecture.[20]

The 1692 edition of the Guizhou Gazetteer shows no appreciable

18. Niu Pinghan, ed., *Qingdai zhengqu yange zongbiao* (Peking: Zhongguo ditu chubanshe, 1990), 365.

19. In spite of the overwhelming evidence for increased geographical familiarity within Guizhou between 1608 and 1673, there is one puzzling aspect as well. Although an analysis of all the place names mentioned in each of the two texts reveals ten new place names appearing in 1673, a dozen names that are mentioned in 1608 do not recur in 1673. Only two of these places are not found on both Ming and Qing historical maps. The remaining ten of these names thus clearly still existed during the Qing. With the exception of Xishan yangdong township and Yongcong to the extreme southeast (Liping prefecture), and Zunyi and Dizhai in the north-central part of the province (Zunyi prefecture), the other place names that do not reappear are in the central part of the province and in relative proximity to other place names mentioned in 1673. Also, with the exception of Zunyi (north-central Guizhou), Zhenyuan (eastern Guizhou), and Yongcong (southeast Guizhou) districts, all of the place names that do not recur in 1673 are of the administrative unit township *(si)*. They may have been small enough that the author of the gazetteer did not mention them for that reason, but instead named larger administrative units.

20. A brief comparison of several entries will serve to illustrate the increased attention to specificity of location. The 1608 entry for the Kemeng Guyang Miao simply states that they live in Jinzhu. The 1673 entry adds the information that Jinzhu had the status of the administrative unit of township, and that it is found in Guangshun department. Similarly, the 1608 entry on the Jiugu Miao states that they are found in Xinglong and Kaili. The gazetteer, however, specifies Xinglong protectorate *(wei)* and Kaili township, revealing once again more attention to administrative detail. The Caijia entry in 1673 specifies that the Caijia are found in Weiqing and Pingyuan; the earlier entry (which included both Songjia and Caijia) did not indicate where either group lived, only that they were originally from the central states *(zhongguo)*.

difference in place names mentioned as compared to the 1673 edition,[21] but the 1741 gazetteer again mentions more place names, and also reveals increased detail regarding place names within the entries and specifies locations for groups where sometimes none were indicated in 1673 or 1692.[22]

The Hei Miao (Black Miao) entry in the 1741 gazetteer provides a good example of how the appearance of a new group in the taxonomy can correspond with territory also mentioned for the first time. The lengthy entry, consisting of thirty-two lines, begins with a listing of the areas where Hei Miao are found: Duyun's Bazhai and Danjiang, Zhenyuan's Qingjiang, and Liping's Guzhou, all place names that did not appear in the earlier texts. Evidently a newfound familiarity with these areas allowed the author to gain information about the people dwelling there.

Other indications of increased geographical connaissance are found in the revised entries on the Zijiang Miao and the Zhongjia. Zijiang Miao have for the first time been identified as living in Pingyue and Guangshun, and these are distinguished from other Zhongjia in certain ways: they can understand Chinese; many are strong and good at war, and some enter the ranks of the army; still others sit for the civil service exams. The revised entry on the Zhongjia also clearly distinguishes between the differing character of subgroups according to the region where they live and the population density of Zhongjia in that area.

By contrast to the consistent increase in the territorial scope of ethnographic inquiry reflected in texts from 1560 through 1741, by the time Li Zongfang compiled his *Qian ji* (1834), the biggest part of the push for territorial expansion and control was over. The significant increase in the number of entries—from thirty-eight to eighty-two—between 1741 and 1834 does not reflect a major increase in familiarity with the territory of Guizhou province.

CREATION OF A COMPLEX TAXONOMY

The problem of classifying and categorizing is at the heart of the Qing effort to understand, represent, and ultimately define the minority groups

21. The only difference is that the Jiantou Gelao are said to be found in Xintian in 1673, and in Guiding in 1692.

22. The 1741 Bai Miao entry mentions Guiding, Longli, and Qianxi zhou; the Qing Miao entry also adds the additional place names of Xiuwen county, Zhenning zhou, and Qianxi zhou; and the 1741 Yangbao entry mentions Zunyi for the first time.

living in Guizhou province. Over the course of more than two centuries the groups identified increased from only thirteen to eighty-two, each associated with its own distinctive practices and characteristics.

The system had its own internal logic. These eighty-two varieties are primarily subdivisions of seven larger groups, namely the Miao, Zhongjia, Gelao, Luoluo, Yao, Zhuang, and Dong, still recognized today. These larger categories (dating from before the Qing) were based on the history, geography, and internal cohesiveness of distinct groups. Other groups fit less nicely into this schema and are apparently more loosely related. These include groups with the suffix "people" *(ren)*, and other assorted individual groups such as the Gedou.

The following examination of the increasingly complex taxonomy of non-Han groups in Guizhou highlights the recurring theme of the need to better understand, more thoroughly identify, and better anticipate the natures and customs of the peoples in question—for administrative purposes. Significant too are the links that remain between the categories ascribed in the Qing and those in use today; the categorization of non-Han peoples on the southwestern frontier was a step on the way to constructing ethnic diversity as a feature that helps to define the People's Republic of China today. Although ethnographers now classify Guizhou's non-Han populations primarily according to language group (a category given little attention under the Qing), as well as their perceived history and geographical provenance, they do not argue with the groupings used in the Qing, but still speak in terms of Miao, Buyi (Zhongjia), Yi (Luoluo), Gelao, Yao, Zhuang, and Dong.

Examining the history of the names of the groups and deducing the basis on which smaller divisions within the larger groups were made sheds light on the growth of the of the Qing system and its internal logic. Table 1 shows the increasing complexity of the taxonomy of groups between 1560 and 1834. Table 2 lists each of the eighty-two names in use by the early nineteenth-century (and probably before), including the basis for that name, and a translation when possible.

Growth of Categories

The first chapter, or *juan*, of Tian Rucheng's 1560 *Yanjiao jiwen* contains information on sixteen different groups, many but not all of whom were resident in Guizhou: Miaoren, Luoluo, Gelao, Mulao, Yanghuang, Zhongjia,[23] Songjia, Longjia, Ranjia, Boren, Yaoren, Zhuangren,

---

23. *Zhong* is written here with the dog radical.

Liaoren, Liren, Danren, and Maren.[24] The entries are brief and served as a source for Guo Zizhang's 1608 *Qian ji*, although six of the groups mentioned in the 1560 work do not recur at all in later ethnographic accounts of Guizhou's inhabitants. The thirteen groups described in Guo Zizhang's 1608 *Qian ji*—Miaoren, Luoluo, Gelao, Yanghuang, Zhongjia, Songjia, Longjia, Si Longjia, Turen, Manren, Dongren, Yangbao, and Boren—therefore represent an increase of five groups recorded for Guizhou province.[25]

Sixty-five years later, the 1673 Guizhou Gazetteer contains a total of thirty illustrated entries on non-Han groups living in Guizhou. The texts are located on one or more openings directly following the illustrations of the group that they describe. The full space allotted for the text is not always filled. Although thirty is a large increase from the thirteen entries of *Qian ji*, the difference is in part accounted for by the fact that, compared with the earlier texts, more subgroups are given their own entries. For example, there are different entries for each of the various kinds of Miao, rather than one entry for Miao with numerous subheadings. Even so, the gazetteer deals with more groups than *Qian ji*. The Hua Miao, Qing Miao, Bai Miao, Hong Miao Sheng Miao, Gulin Miao, Yangdong Luohan Miao, and Mulao are all new in 1673.

Other changes in organization and presentation regarding groups that appear in both texts are as follows: the Luoluo now appear under two headings, Hei and Bai (Black and White). The Gelao appear under two headings, Daya Gelao and Jiantou Gelao, although all five kinds mentioned in the single *Qian ji* entry (Hong, Hua, Daya, Jiantou, and Zhushi) are still discussed within these parameters. The Songjia and Caijia, who earlier appeared together under the Songjia heading, each have their own entries. The Bafanzi, who were discussed under the heading of "Miaoren," now have their own entry. The Pingfa si Miao,

24. The meaning of the names of the groups is discussed in the section on naming below. Subgroups are mentioned for the Miao, Luoluo, and Gelao as follows: for the Miao, Kemeng and Guyang Miao (noted as two kinds, but in a single sentence), Jiuming Jiuxing Miao, Zijiang Miao, Maiye Miao, Duanqun Miao, Bafanzi, and Hei or Yao Miao; for the Luoluo, Hei and Bai (also called Wu Man and Bai Man respectively); and for the Gelao, Hua Gelao, Hong Gelao, Daya Gelao, Jiantou Gelao, and Zhushi Gelao.

25. Subgroups are as follows: for the Miao, Dong Miao Xi Miao, Bafanzi, Kemeng Guyang, Maiye Miao, Duanqun Miao, Dong (east) Miao, Xi (west) Miao, Jiuming Jiuxing Miao, Zijiang Miao, Li or Yao Miao, Duanqun Miao, Jiugu Miao, and other Miao groups identified only by place name; the Luoluo again have two subdivisions, Hei and Bai; Gelao subgroups, similarly, are once again Hua Gelao, Hong Gelao, Daya Gelao, Jiantou Gelao, and Zhushi Gelao.

mentioned only in passing in *Qian ji* (1608), here receive their own entry. The remaining groups from *Qian ji*, Turen, Yanghuang, Manren, Yangbao, Boren, and Dongren, each maintain their own entries as in the 1673 gazetteer.

As when comparing Tian Rucheng's 1560 text with Guo's 1608 text, it is not simply a matter of more groups being added to an already existing database, so to speak. Rather, as more names appear, others are dropped in a process of refinement. The Maiye Miao and Jiuming Jiuxing Miao (both listed under the Miaoren entry in 1608) drop out. The Longjia and Si Longjia also fail to appear in the 1673 gazetteer, but much of the same information is listed under the headings of Gouer and Madeng Longjia, new in 1673. The broad categories of *sheng* ("wild" or "raw") and *shu* ("tame" or "cooked") Miao, used to speak of the Miao in general terms in *Qian ji*, are also absent from the entries of the more systematized 1673 gazetteer. Its finer distinctions between groups render this broadest level of categorization outmoded. The gazetteer's increasing refinement and systematization in categorizing and naming non-Han groups residing in Guizhou province is significant.

Compared with the 1673 Guizhou Gazetteer, only one new entry is included in the 1692 edition, that of "Nüguan"—the title of the primary wife of the Luoluo chieftain.

The 1741 Guizhou Gazetteer again shows a significant increase in the number of groups that receive entries. It contains thirty-eight as compared to the 1692 Gazetteer's thirty-one. The entries are similar in general style and content to the texts introduced above, but they are generally more extensive. The increase in number does not reflect the straightforward addition of seven new groups to the classification system, but rather a continual process of refinement and adjustment in the naming and categorizing Guizhou's non-Han population. Several of the category headings included in 1673 and 1692 do not reappear in 1741 (Duanqun Miao, Jiantou Gelao, Gouer Longjia, Madeng Longjia). The organization of the names, and the order in which they appear, has also undergone additional change. (See table 1 for a full list of entries in each work discussed).

Between the 1741 edition of the Guizhou Gazetteer and the appearance of Li Zongfang's *Qian ji* (prefaced 1834), the number of groups that have their own entries has more than doubled, from thirty-eight to eighty-two (the number of groups most often included in the Miao albums). This increase simply reflects a further division of existing groups into subgroups with their own entries. For example, the Hei Miao now appear under six headings: Hei Miao, Bazhai Hei Miao, Qingjiang Hei Miao, Louju Hei Miao, Hei Shan Miao, and Hei Sheng Miao. No new groups have been added to the taxonomy.

Table 1. Entry Headings for Non-Han Groups Mentioned in the 1560, 1608, 1673, 1741, and 1834 Texts. Between 1560 and 1834 the number of entries found in the texts under discussion grew from sixteen in 1560 (and only thirteen in 1608) to eighty-two in 1834. The more intimately they could be described and recognized, the more easily appropriate techniques of control for each group could be devised. The following chart reveals at a glance the headings for the entries each of the six texts contained and how the number grew over time.[26]

Group Names	1560	1608	1673	1741	1834
Bafan Miao			×	×	×
Bai Erzi					×
Bai Miao			×	×	×
Bai Longjia					×
Bai Luoluo			×	×	×
Bai Zhongjia					×
Bazhai Hei Miao					×
Boren	×	×	×	×	×
Bulong Zhongjia					×
Caijia			×	×	×
Cengzhu Longjia					×
Chezhai Miao					×
Danren	×				
Datou Longjia					×
Daya Gelao			×	×	×
Dong (cave) Miao					×
Dong (east) Miao				×	×
Dong Miao ×i Miao			×		
Dongjia Miao					×
Dongren		×	×	×	×
Dongzai Miao					×
Duanqun Miao			×		×
Gaopo Miao					×
Gedou				×	
Gelao	×	×		×	
Gezhuang Miao					×

(Continued on next page)

26. The chart must be used with care because it is based entirely on the names of entries as they appear in each text. Thus a cursory glance at the "Luoluo" row reveals no entry for 1673/90. This does not mean that the Luoluo do not appear at all, but rather that they are found under two headings, Bai Luoluo and Hei Luoluo. In 1741 they appeared under the entries Luoluo and Bai Luoluo, in 1834 under Luoluo, Bai Luoluo, and Luogui Nuguan. Similar phenomena occur under the Gelao, Miaoren, Longjia, Sheng Miao Hong Miao, and Jiugu Hei Miao entries, where the categories at first appear to drop out, but closer inspection reveals the group is actually listed under several subheadings.

Table 1. *Continued*

Group Names	1560	1608	1673	1741	1834
Gouer Longjia			×		×
Gulin Miao			×	×	×
Guoquan Gelao				×	×
Guyang Miao			×		
Hei Luoluo			×		
Hei Miao				×	×
Hei Zhongjia					×
Hei Jiao Miao					×
Hei Lou Miao					×
Hei Shan Miao					×
Hei Sheng Miao					×
Hong Gelao					×
Hong Miao				×	×
Hongzhou Miao					×
Hua Gelao					×
Hua Miao			×	×	×
Hulu Miao					×
Jianding Miao					×
Jianfa Gelao					×
Jiantou Gelao			×		
Jiugu Hei Miao			×		
Jiugu Miao				×	×
Jiuming Jiuxing Miao					×
Kayou Zhongjia					×
Kemeng Guyang Miao				×	×
Langci Miao					×
Liao ren	×				
Liminzi					×
Linjia Miao					×
Liren	×				
Liu Ezi				×	×
Liudong Yiren					×
Longjia	×	×		×	
Louju Hei Miao					×
Luogui Nüguan					× [27]
Luohan Miao					×
Luoluo	×	×		×	×
Madeng Longjia			×		×
Maren	×				
Manren		×	×	×	×
Miaoren	×	×			
Mulao	×		×	×	×
Nong Miao				×	×
Pingfa(si) Miao			×	×	×
Pipao Gelao				×	×

(*Continued on next page*)

27. Same as Nüguan or Naide.

Table 1. *Continued*

Group Names	1560	1608	1673	1741	1834
Qian Miao					X
Qing Miao			X	X	X
Qing Zhongjia					X
Qingjiang Hei Miao					X
Qingjiang Zhongjia					X
Ranjia	X				
Ranjia Man					X
Sheng Miao					X
Sheng Miao Hong Miao			X		
Shui Gelao				X	X
The Six Types: Shui, Yang, Ling, Dong, Yao, Zhuang				X	
Shuijia Miao					X
Si Longjia		X			
Songjia	X	X	X	X	X
Tu Gelao					X
Turen		X	X	X	X
Xi Miao				X	X
Xiqi Miao					X
Yangbao		X	X	X	X
Yangdong Luohan Miao			X	X	X
Yangguang Miao					X [28]
Yanghuang	X	X	X	X	
Yao Miao			X	X	X
Yaoren	X			X	X
Yaque Miao					X
Yetou Miao					X
Zhongjia	X	X	X	X	
Zhuangren	X				
Zhushi Gelao					X
Zijiang Miao			X	X	X

Basis for Naming

The names used in local gazetteers and Miao albums to distinguish sub-groups combined designations that indicated to what broader ethnic group they belonged (based on the history, geography, and internal cohesiveness) with specific observable traits (i.e., dress, customs, location, livelihood, dwelling type, and nature of individual groups). Apart from the fact that many of the group names were written with the dog radical, few of the names themselves were derogatory. Epithets such as "Pig Filth Gelao," were an exception to the norm. More often, groups were named for relatively neutral criteria such as the color of their clothing.

28. What had been written as Yanghuang shifted in orthography to become Yangguang.

In some instances the derivation of names is unclear. In any case, it was a system whereby an outsider, looking in, categorized peoples according to traits easily observable from outside. These Qing dynasty labels are instructive when it comes to understanding the taxonomy that evolved for Guizhou's non-Han peoples, and in revealing attitudes that the Chinese held toward others.

Miao. During the Qing, Guizhou was home for more subgroups of Miao than for any other group. By the turn of the nineteenth century no fewer than forty-four varieties of Miao had been enumerated (see table 1). These subgroups of Miao were labeled in various ways according to the characteristics of their dress, peculiarities of custom, distinctive features of their dwelling places, or other characteristics remarkable to their observers (see table 2).

Miao who were classified according to the color of their dress are the Hua (flowery), Hong (red), Bai (white), Qing (blue), and Hei (black) Miao. Miao named according to location are the Dong (east) and Xi (west) Miao, Qingjiang (Qing river) Miao, Pingfa Miao, Hongzhou Miao, Xiqi (west of the river) Miao, Chezhai Miao, and Yao Miao. According to Beauclair, "the name Yao Miao derives from an administrative district around [Duyun], which was ruled by a Yao family from [Guangxi] for 500 years."[29] The Nong (agricultural) Miao and Shuijia (water-*jia*) Miao are named for their occupations. Dong (cave) Miao, Gaopo (high-cliff) Miao, Qian (dense bamboo grove) Miao, and Louju (storied-house-dwelling) Miao are named for their dwelling places; Yaque (sparrow) Miao for the sound of their speech. The Duanqun (short-skirt) Miao are named for their costume, and the Jianding (pointed-head) Miao for their hairstyle.[30]

Some groups have compound names that seem to derive from several traits. This can be seen in the case of the various kinds of Hei (black) Miao. For example there are Hei Shan (mountain) Miao, Hei Sheng (wild) Miao, Hei Lou (storied-house) Miao, and Hei Jiao (foot) Miao. Thus, various kinds of Hei Miao are further subdivided to distinguish between them.[31]

---

29. Beauclair, *Tribal Cultures of Southwest China*, 121.

30. Names for subgroups of Miao are also discussed in Ramsey, *Languages of China*, 281.

31. The 1741 Guizhou gazetteer subdivides the Hei Miao as follows: those who dwell in mountainous areas are called Mountain *(shan)* Miao or Gaopo (high-terrace) Miao; those who live near rivers are called Dong (cave) Miao; those who are under the *tusi* system are called Shu (tame) Miao; those who have no officials presiding over them are called Sheng (wild) Miao *(wu guan zhi wei sheng miao).* 7:14a.

Other kinds of Miao are named for their nature—for example, the Sheng (wild) Miao—or for a distinctive aspect of their history. The Jiugu (nine-divisions) Miao are said to be the descendants of the only nine people (of the tribe) who survived Zhuge Liang's southern campaigns during the later Han dynasty.[32] The Yanghuang were descended from the Yang family who had lived in Bozhou.[33] The Zijiang Miao had been under the magistrate of Zijiang during the Tang dynasty; their name derives from the location where they lived.[34]

Luohan Miao is a good example of a name that has been endowed with various interpretations over the course of time. According to Beauclair, "The designation [Luohan] for young tribal people occurs repeatedly in the Chinese records, and must be a phonetic rendering of an expression in a non-Chinese language, probably that of the Yao." One Chinese source describes them as "bad young people," another relates that "the Yao call their young men over eighteen years [Luohan]."[35] The makers of the Miao albums based their illustrations on an interpretation of Luohan as meaning "arhat"; a statue of the Buddha with worshippers in front of it almost always forms the content of the accompanying illustrations. These two different interpretations may not be mutually exclusive. In Hinayana Buddhist societies such as Thailand, virtually all young men were expected to become monks for a time after they attained maturity. The Tai of Sipsuang-banna (in Yunnan) and the Zhuang of Guangxi are very closely linked to the Thai people. The Luohan Miao of Guizhou may have been of similar extraction.

The origins of the names of other Miao subgroups are not as clear. They may be transliterations of sounds from native languages into Chinese, or stem from some other now unknown source.

Zhongjia. Beginning with the 1741 gazetteer, three types of Zhongjia are enumerated (although they receive individual entries only later). These are the Kayou, Bulong, and Qing Zhongjia. The Qing Zhongjia are named for the color of their costume. The Bulong may be named after the surname of their leader.[36] The derivation of the name Kayou is unclear.

Gelao. There are as many as eight subgroups of Gelao. The Hua Gelao and Hong Gelao are named for the color of their garments; the

---

32. From the Jiugu Miao entry in the *Huangqing zhigong tu*. See Chuang, *Xie Sui*, 575.
33. Lin Yüeh-hwa, "The Miao-man Peoples of Kweichow," 278.
34. Ibid., 280.
35. Beauclair, *Tribal Cultures of Southwest China*, 120, note 1.
36. Clarke, *Among the Tribes of Southwest China*, 84.

Pipao, or "cape," Gelao for their capes; Jiantou (pointed-head) Gelao for their hairstyle. The Daya Gelao are named for their conspicuous custom of knocking out or breaking off a bride's teeth before she moves in with her husband's family. The Shui Gelao are named for their skill in the water and the fact that their livelihood is largely derived through fishing. The Zhushi (pig-filth) Gelao are disliked because of their fierce and rude nature and their belligerent reputation. Album texts state that they bathe only once a year, which is probably how they got their name. The Tu (local) Gelao are distinctive in the albums for the grass cloaks they wear and the procedure they use to toughen their feet. The derivation of their name is unclear.

Luoluo. Two groups of Luoluo are commonly referred to in eighteenth-century ethnographic texts. They are the Hei, or "black," Luoluo, and Bai, or "white," Luoluo. Miao albums often refer to the Black Luoluo as the aristocracy *(da xing),* and to the White Luoluo as their subordinates or even slaves *(xia xing).*[37] Victor Mair has found that the White Luoluo are "generally the descendants of Han Chinese slaves who have been enculturated by their former masters in spite of the fact that they outnumber them by a huge majority."[38] A third entry often appearing in album entries under the heading "Nüguan" or "Naide" is devoted to the first wife of the Black Luoluo chieftain, who had the power to act as regent for her son in case her husband should die.

Yao. According to the albums, eighteenth-century Guizhou had only one kind of Yao, called "Yaoren."[39] In areas where the Yao were more numerous, they were most often categorized into subgroups according to the regions where they lived.

Longjia. The Longjia frequently appear in the Miao albums and in local gazetteers. By 1834 they were considered to be comprised of a total of five subgroups. Four of the five different subgroups of Longjia were named for some aspect of their dress. The appellations Madeng (stirrup), Gouer (dog-ear), and Datou (large-head) Longjia all refer to the hairstyle of the women. Bai (white) Longjia are named for their white clothes. The derivation of the term Cengzhu Longjia is unclear.

The basis on which divisions among the subgroups were made during the Qing seems problematic and artificial from a twentieth-century

---

37. See Ramsey, *Languages of China,* 252–53 on class divisions among the Luoluo.
38. Mair, "The Book of Good Deeds," 405.
39. The Yao Miao referred to above were treated as a kind of Miao. The character for "Yao" in the compound Yao Miao is different from that for Yaoren.

vantage point because the names were based on externally observable (or ascribable) characteristics rather than on distinctions articulated from within the communities in question. However, even if externally imposed, these categories were not without meaning. Richard Cushman, who has worked on what he calls the ethnohistory of the Yao, is of the opinion that, although various subgroup names denoting the Yao have been used for centuries, "no consistent over-all typology ever existed among the Chinese."[40] He believes rather that these subgroup names reflect the "illusions of Chinese cross-cultural perception," and that any such divisions are false, or nonexistent. He draws the conclusion that "[t]here seems to be no reason, then, given the present state of knowledge of Yao culture, to believe in the existence of subgroups or tribes among the Yao in China."[41]

While it is certainly true that not only Yao but also Miao, Zhongjia, and other subgroups were labeled and thus in a sense "created" by the Chinese, to deny the existence of these subgroups is not particularly helpful. Cushman recognizes the possibility that the subgroup names were in use for long enough by the Chinese that some Yao groups may eventually have accepted and even adopted them. This may even be the reason, he asserts, that a handful of Yao groups are individually endogamous.[42] Thus artificial categories may have been imposed on the Yao by the culturally dominant Chinese, but those categories certainly did *exist*. Cushman's argument, that certain groups of Yao may have become individually endogamous because of the categories the Chinese imposed on them, attests to the power of these categories and their ultimate fundamental influence on Yao culture.[43] If Chinese categories did indeed

40. Richard David Cushman, "Rebel Haunts and Lotus Huts: Problems in the Ethnohistory of the Yao" (Ph.D. diss., Cornell University, 1970), 91.

41. Ibid., 98.

42. Ibid., 99, note 19.

43. I am not entirely convinced that the categories into which the Chinese divided the Yao subgroups had an effect on their marriage patterns. Cushman's logic seems to be that the categorization of groups according to superficial differences, i.e., color of clothing, or location where a group lives, are simply superficial. I would argue that such apparently superficial distinctions, while not significant in themselves, may be an outward manifestation of differences that go deeper. To take a current and more local example we can look at different groups of North American Amish. Within the Amish there are many subgroups. Epithets include, for example, "yellow toppers" (for the yellow tops on this group's buggies) and "Nebraska" (the "plainest" group). The names of the different groups describe only outward appearances or places, but the differences between the groups are quite significant and include questions of belief and lifestyle. In this case at least it is clear that the differences existed before the names. It is likely that they were then named according to outwardly apparent differences by outsiders who did not grasp the full significance of the distinctions between the groups.

affect Yao behavior and perceptions of self, categories ascribed to other groups may also have affected the lives of Miao, Luoluo, Zhongjia, and other groups—even if only with regard to the ways in which Chinese officialdom related to them.

The system of classification used in the eighteenth century had its own purpose and its own internal logic. To deny it validity only because it does not correspond to our own present-day criteria is to lose sight of what earlier sources can tell us about their creators, their uses, and the era from which they stemmed.

Table 2. Eighty-two Most Commonly Appearing Names of Non-Han Groups Residing in Guizhou. Derivations and translations appear alphabetically by group (Miao, Luoluo, etc.).

Name	Basis for Name	Translation or Transliteration of Name
Bafan Miao	?	Bafan Miao
Bai Miao	dress	White Miao
Chezhai Miao	location	Chezhai Miao
Dong (east) Miao	location	East Miao
Dong Miao	dwelling place	Cave Miao
Dongjia Miao	family name ?	Dongjia Miao
Dongzai Miao	?	Dongzai Miao
Duanqun Miao	dress	Short-Skirt Miao
Gaopo Miao	dwelling place	Tall-Cliff Miao
Gulin Miao	?	Gulin Miao
Hei Miao	dress	Black Miao
Hei Shan Miao	dress, dwelling place	Black Mountain Miao
Hei Sheng Miao	dress, nature	Black Wild Miao
Hei Jiao Miao	dress and ?	Black-Foot Miao
Hei Lou Miao	dress and dwelling	Storied-House Black Miao
Hong Miao	dress	Red Miao
Hongzhou Miao	location	Hongzhou Miao
Hua Miao	dress	Flowery Miao
Hulu Miao	?	Gourd Miao
Jianding Miao	hairstyle	Pointed-Head Miao
Jiugu Miao	history/legend	Nine-Divisions Miao
Jiuming Jiuxing Miao	nature/disposition	Nine-Names and Nine-Surnames Miao
Kemeng Guyang Miao	?	Kemeng Guyang Miao
Langci Miao	custom (couvade)	Virtuous and Kind Miao [44]
Lingjia Miao	?	Lingjia Miao
Louju Miao	dwelling place	Storied-House-Dwelling Miao
Luohan Miao	(see text)	Arhat Miao
Nong Miao	name or occupation	Agricultural Miao
Pingfa Miao	location	Pingfa Miao

(*Continued on next page*)

44. *Langci* is an epithet for "mothers."

Table 2. *Continued*

Name	Basis for Name	Translation or Transliteration of Name
Qian Miao	dwelling place	Dense-Bamboo Miao
Qing Miao	dress	Blue Miao
Qingjiang Miao	location	Qing (clear) River Miao
Sheng Miao	nature/disposition	Wild Miao
Shuijia Miao	occupation	Water "jia" Miao
Xi (west) Miao	location	West Miao
Xiqi Miao	location	West Stream Miao
Yangbao Miao	?	(Willow-Protecting?) Miao
Yangdong Luohan Miao	?	Yangdong Luohan Miao
Yanghuang Miao	family name [45]	Yanghuang Miao
Yao Miao	location	Yao Miao
Yaque Miao	sound of speech	Sparrow Miao
Yetou Miao	?	Elderly-Headman Miao [46]
Zijiang Miao	location	Purple-Ginger Miao
Bai Luoluo	?	White Luoluo
(Hei) Luoluo	?	(Black) Luoluo
Nüguan	custom	Female Official
Caijia	family name ?	Caijia
Songjia	family name ?	Songjia
Qing Zhongjia	dress	Blue Zhongjia
Bai Zhongjia	dress	White Zhongjia
Bulong Zhongjia	family name	Basket-Repairing Zhongjia [47]
Hei Zhongjia	dress	Black Zhongjia
Kayou Zhongjia	?	Kayou Zhongjia
Qingjiang Zhongjia	location	Clear River Zhongjia
Bai Longjia	dress	White Longjia
Cengzhu Longjia	?	Cengzhu Longjia
Datou Longjia	hairstyle	Large-Headed Longjia
Gouer Longjia	hairstyle	Dog-Eared Longjia
Madeng Longjia	hairstyle	Stirrup Longjia
Daya Gelao	custom	Teeth-Breaking Gelao
Guoquan Gelao	hairstyle	Pot-Ring Gelao
Hong Gelao	dress	Red Gelao
Hua Gelao	dress	Flowery Gelao
Jiantou Gelao [48]	hairstyle	Haircutting Gelao
Pipao Gelao	dress	Cape Gelao

(*Continued on next page*)

45. Lin Yüeh-hwa, "The Miao-man Peoples of Kweichow," 278.

46. Ruey Yih-fu translated this as "great stockade-living Miao." See the table of contents to Ruey Yih-fu, ed., *Miao luan (man) tu ce* (Taipei: Academia Sinica, Institute of History and Philology, 1973), 2.

47. This translation is given by Ruey, *Miao luan (man) tu ce*, table of contents.

48. Sometimes called Jianfa Gelao.

Table 2. *Continued*

Name	Basis for Name	Translation or Transliteration of Name
Shui Gelao	custom (livelihood)	Water Gelao
Tu Gelao	?	Local Gelao
Zhushi Gelao	nature/disposition	Pig-Filth Gelao
Boren	dwelling[49]	Boren
Dongren	dwelling	Cave People
Liudong Yiren	dwelling	Six-Cave Yi (barbarian) People
Manren	?	Barbarian People
Turen	?	Local People
Yaoren	?	Yaoren
Mulao	(see text)	Wood Rats
Ge Zhuang	?	Wild Dog
Liu Ezi	?	Six Ezi
Bai Ezi	?	White Ezi
Ranjia Man	?	Ranjia Man
Liminzi	?	Liminzi
Bai Erzi	custom (intermarriage)	White Sons

THE SHIFT TO DIRECT OBSERVATION

When new groups are added to the taxonomy, or entries on existing groups are revised, the nature of the content points to an emphasis on direct observation in describing or reporting on the non-Han residents of Guizhou. Examination of *Qian ji* (1608), for example, demonstrates that additional research had been carried out on groups previously mentioned by Tian Rucheng in 1560. While entries appearing in both texts are similar in content—the 1560 texts having formed the basis for much of what is recorded in 1608—the later entry has often been expanded. The 1608 Luoluo entry contains almost a full page of additional information. This includes what the Luoluo do when someone falls ill—namely calling on a male shaman rather than relying on doctors and medicine—and a section on the chieftain's wife. In the case of the Gelao, the 1608 entry includes several lines more than in 1560. Guo informs his reader that the Gelao in Qingping understand Chinese[50] and

49. According to Lin Yüeh-hwa, "The Miao-man Peoples of Kweichow," 280, note 58, the character for Boren was made by placing the characters for "brambles" with that of "person." They were named for a thorny, inhospitable environment.

50. *Hanyu* translates literally as "language[s] of the Han."

that they observe the laws of the government *(ting guanfu yueshu).*[51] Additionally he mentions peoples in Shiqian's Miaomin township and Liping's Badan, Guzhou, and Caodi townships whose customs are similar to the Gelao. The customs of those under the jurisdiction of Duyun's Bangshui township are also, he reports, the same as those of the Gelao. The mention of these specific place names suggests increased familiarity on the part of Qing officialdom with the region, and with distinctions (or similarities) among the peoples dwelling there.

The only case where the content of an entry differs completely between the 1560 and 1608 texts is in the descriptions of the Boren. Tian's 1560 entry begins historically, naming the locations in which Boren were found during the Han (206 B.C. to 219 A.D.), Tang (618–906), and Nan Zhao (c. 730–902)[52] periods, and traces the origins of the group. It also mentions that they are Buddhist and use prayer beads, an image that recurs pictorially in Miao albums.

Guo's 1608 entry does not trace the history of the group but shows more interest in its present circumstances. He states that

> all of the commanders of the twelve Boren camps *(ying)* are, in fact, from Luoluo tribes *(buluo).* The languages of the Zhongjia, Gelao, and Boren are not mutually intelligible; the Boren often act as interpreters. Their disposition is fierce *(han)*—violent and tough—and they like to fight. They are like the Luoluo [under the] pacification office *(xuanweisi Luoluo)* in their clothing and dwellings, marriages and funerals, and general tastes *(changhao).*[53]

Although the entries differ considerably, they may refer to the same group since bits of information from both accounts appear in later albums. The shift in the kind of content between 1560 and 1608 is significant, because it reveals that the 1608 text may have been based on observation. The kind of information it contains would have been observable in the ethnographic present whereas the older text probably relied on historical, written, source material.

Some important inferences regarding the growing importance of direct observation can also be made from a comparison of the 1608 and 1673 texts. This is true both in cases of revised and newly added entries.

---

51. Guo Zizhang, *Qian ji* (1608), 59:8a.

52. According to Charles Backus in *The Nan-Chao Kingdom and T'ang China's Southwestern Frontier* (Cambridge and New York: Cambridge University Press, 1981), the Nan Zhao kingdom was consolidated between 730 and 790, and was succeeded by a series of other kingdoms beginning in 902. Some Chinese texts use the term Nan Zhao more broadly to refer to the area now generally encompassing Yunnan until up to the Yuan dynasty (1280–1368) conquest of Yunnan.

53. *Qian ji* (1608), 59:12, 14.

Revisions to the Jiugu, Zijiang, and Zhongjia entries are illustrative. We learn from the 1608 Jiugu entry that this group was not pacified until 1600. This fact may help to explain the nature of the information recorded in the same text; up until the Jiugu Miao's recent submission, groups in bordering areas engaged in revolts would allegedly entice the Jiugu Miao to assist them. Guo records that since their submission, seventy-two *zhai* pay a total of twenty-three piculs of grain in tax per year.[54] Remarkably absent from this earlier entry is the kind of ethnographic information that appears in the 1673 gazetteer, which states that they are similar to Bianqiao's Hei Miao, are fond of the color *qing* (blue), and live in deep caves. All that *Qian ji* says regarding their customs is that they are like those of the Miao in Zhenyuan (who are not further discussed in *Qian ji*!). In other words earlier officials could not get close enough to gather information based on observation. They only knew into what administrative category the Jiugu Miao fell.

As in the case of the Jiugu Miao, the 1608 passage on the Zijiang Miao reveals a proximity to a time when the group was completely outside the ken, or at least the control, of the imperial government. The authors feared the Zijiang Miao: "They are also called Zihong.[55] Their delight in killing is extremely keen. When they get an enemy they chew his raw flesh. When a man dies he is buried only after his wife remarries. This is called 'mourning has a leader.'"[56] The Chinese took an interest in marriage procedures and death rituals as markers of culture. There were proper and civilized ways to carry out certain rites. Conformity with the rites was therefore of significance on the scale of acculturation and status. The remarriage of a widow before her previous husband's burial may be an accurate reporting of customs, or may be a way of conveying the "wild" or uncivilized nature of the Zijiang Miao. Although the tone of the 1673 gazetteer is also critical the entry is less sensational, mentioning neither cannibalism nor the swift remarriage of widows. Rather there is more information comparing them with other groups, and about seasonal customs. The 1673 text shows a shift in content to information that could only be learned as a better knowledge of the group was obtained through observation. In addition, the harshest (probably unverifiable) statements from 1608 do not reappear.

More knowledge does not always mean that there are good relations

54. Ibid., 59:7.
55. Literally "violet red."
56. *Qian ji* (1608), 59:5.

between the group in question and those describing it. Nor does it mean that all of the information would be considered objective or rational by today's standards. The Zhongjia entry is the only one in 1673 that mentions *gu* poison. *Gu* was thought to be used as a poison by its practitioners, who gained wealth and power through its use. The *gu* was actually a kind of animal. To obtain or capture a *gu* the practitioner would put several of various kinds of crawling creatures in a container: spiders, centipedes, etc. They would be left for a period of time. At the end the only creature that was left, having devoured the others, was considered to be the *gu*. The *gu* could then be used to perform spells on the unsuspecting. *Gu* has remained alive and well in the popular imagination. During the early twentieth century, trials for its use were conducted using evidence as presented in a court of law. Even as of the 1980s the belief among the Han that various kinds of Miao are practitioners had not been not wiped out.[57]

In any case, the Han feared the Zhongjia. As the text of the gazetteer continues, we learn that

> [t]heir nature *(xing)* is dangerous and wily, and they are fond of killing. When they go out they must carry a strong crossbow and a sharp knife. If an enemy [so much as] stares, they must take revenge.[58]

The text then proceeds to describe their involvement in banditry and highway robbery. In closing, the entry mentions that in Ming times attacks were repeatedly launched on the intransigent Zhongjia, but whenever the soldiers arrived, the wily natives would disperse. When the soldiers left, they would once again come together, so it was difficult to restrain them. The final lines assure the reader that more recently a measure of success had been achieved against the Zhongjia, that now gradually they would rather put away their weapons. The above analysis shows that the basis for the new entries in the 1673 gazetteer, with the possible exception of the use of *gu*, derives from information obtained through direct observation. To illustrate this point more graphically, table 3 charts new entries against various categories of topics mentioned in the text.

A glance at table 3 reveals that the most consistently reported

---

57. Norma Diamond, "The Miao and Poison: Interaction on China's Southwest Frontier," *Ethnology* 27 (1988): 1–25. For more on *gu* poison, see also H. Y. Feng and J. K. Shryock, "The Black Magic Known in China as Ku," *Journal of the American Oriental Society* 55 (1935): 1–30.

58. *GZTZ* (1673), *juan* 29, *miao liao* subsection, entry no. 19.

Table 3. Non-Han Groups Mentioned for the First Time in the 1673 Guizhou Gazetteer with Topics Appearing in Descriptive Texts.

	1	2	3	4	5	6	7	8	9	10	11	12	13
location	X	X	X	X	X	X	X	X	X	X	X	X	X
appearance	X	X	X	X		X		X	X		X	X	
marriage						X	X					X	
death				X							X	X	
weapons		X			X		X						
their nature *(xing)*	X	X	X	X	X	X	X	X		X			X
livelihood			X	X		X							
religion				X			X				X		X
surnames				X									
diet				X									
other					X*					X†	X‡		

* Active in kidnapping for ransom.
† Mentions dwelling type.
‡ "Their taste for killing is extremely keen."

Key to numbers:

1	Hua Miao	5	Gulin Miao	9	Duanqun Miao
2	Qing Miao	6	Yangdong Luohan Miao	10	Guyang Miao
3	Bai Miao	7	Mulao	11	Pingfa si Miao
4	Sheng Miao Hong Miao	8	Dong Miao Xi Miao	12	Yao Miao
				13	Zijiang Miao

types of information are the location in which a certain group lived and the appearance of its members—both ascertainable through direct observation—and something about their "nature." Interestingly, of the four groups whose appearance or costume is not remarked upon, three are noted either as having dispositions that are fierce, sly, deceitful, and/ or gluttonous, and two of the entries comment on what kinds of weapons they use or carry. It would seem that the "tame" groups were more approachable and possibly more easily observable in the marketplace, whereas it would have been more difficult to get a good look at those who were considered to be more fierce.

The 1673 Guizhou Gazetteer pushes ethnographic representation to a new level by including illustrations. How true-to-life the illustrations were, however, is somewhat questionable. Each illustration fills an open spread (two leaves). Generally speaking, each illustration shows three to four figures in a natural setting. The figures are always approximately the same size, and always have the same placement on the page. They are roughly centered between the top and bottom margins; their height is not quite equivalent to half the total height of the page. They do not look like distinct individuals—their features are mostly nondescript—but function rather as props on which to hang clothing. The costumes shown, and activities engaged in, vary according to the group. The women tend to be dressed more elaborately earlier in the sequence where, for example, they often sport ornate vests. Later in the sequence this degree of intricate detail tapers off, as if the "best" illustrations were put up front to impress the viewer, and toward the end of their task the artists simply tired or needed to hurry in order to meet a deadline, and thus gave less attention to detail and ornamentation. Some of the illustrations that appear later in the series are almost certainly not drawn from observation, or even textual accounts; some of the entries include no discussion of costume. As with depictions of Columbus's voyages that included illustrations for general interest, accuracy was questionable at best. By contrast, the illustrations in the 1692 Guizhou Gazetteer appear to be more individualized, and as such they may be more objectively representative. As mentioned above, these later illustrations are similar in both content and composition to the Miao albums, topic of the next chapter, and may have been the basis on which many of the illustrations in the albums were made.[59]

The texts of the 1673 and 1692 Guizhou Gazetteers also show some differences, although they are generally fairly minor in nature. The most new information is recorded about the Kemeng Guyang Miao:

> They use a tall ladder to go up and down [from their dwelling places]. In cultivating the earth they do not use a plough, but instead a small hoe. They rake the earth [after sowing grain] but do not weed. Men and women play the *lusheng* and pair off. If no child is born the betrothal gifts are returned. When a parent dies they do not mourn. They laugh and dance, and sing with gusto. They call this making noise for the corpse. The next year when they hear the cry of the cuckoo reach their house they weep silently. The bird [returns] like the [new] year; but the loved one will not come back.[60]

Again, one can only surmise that this information is the result of some kind of fieldwork.

We saw above that the 1741 Guizhou Gazetteer contains more entries than the 1673 or 1692 editions. In addition they are generally more extensive. Although some text appears to be borrowed from the earlier gazetteer, many of the entries contain additional or entirely new

59. New illustrations continued to be made during the eighteenth century; many of the Miao albums have eighty-two entries, whereas the 1692 gazetteer contained only thirty-one.
60. *GZTZ* (1692), 31:58b.

information, pointing toward a continued effort at collecting and refining information.

The content of the Hei Miao entry supports the argument that direct observation played an increasingly important role in information gathering. The second part is organized according to different locations (all mentioned for the first time in 1741) and explains how the customs of those in Danjiang differ from those in Bazhai, what is characteristic of those in Qingjiang, what is distinctive about those in Guzhou and in Dongzai. Although we have no clues beyond the text itself, from the structure of the account it would seem as if the author had traveled to these different areas and compared them in his notes, from which he then wrote up such an entry. It is also possible that reports had been submitted from these various regions that were then compiled by the author. In any case the information does not seem to have been compiled from earlier publications.

The entries in Li Zongfang's *Qian ji,* prefaced 1834, although more numerous than in the other texts discussed, tend to be significantly shorter, and do not reflect an increase in direct observation in the same way as the earlier texts. In fact, Li Zongfang condensed the entries that appeared in an earlier Miao album. *Qian ji* represents a falling off in the interest of collecting as much information about each group as possible. Rather the entries here contain so little information they can be thought of as captions (to nonexistent or missing illustrations) rather than freestanding essays.

ADMINISTRATIVE GOALS ACCOMPLISHED

This section concludes the discussion of the increasingly ethnographic nature of these texts by considering their attention to questions relating to administration of Guizhou's non-Han groups, and particularly to their "taming." Comments on the degree of sinicization of various groups appear with some frequency in the first Kangxi Guizhou Gazetteer. The 1673 entry on the Guyang Miao shows evidence of categorizing groups according to their degree of docility and/or acculturation. The entry ends with the following statement:

> Of all the Miao, only the first four kinds[61] are pure and simple *(chunpu).* They shrink from seeing officials. When there is an injustice they comply with what

61. The author refers to the four types listed in the first three entries—Hua Miao, Dong Miao Xi Miao, and Kemeng and Guyang Miao. The Kemeng and Guyang Miao are considered to be

their village elder decides. They pay rent and provide services like loyal subjects, therefore their poverty is especially keen.[62]

This difference in the 1673 text over earlier texts shows a direct concern with matters concerning administration and governance. Another instance of commenting on acculturation is found in both the 1673 and 1692 Songjia entries. A final line states: "Now they have moreover changed completely and become Chinese, and do not revert [to their old ways]."[63] The general tone of the comment seems to be more one of satisfaction than of regret.

In the 1741 gazetteer, too, we see an increasing number of comments regarding the sinicization of particular groups. Of Pingyue and Guangshun's Zijiang Miao it says: "If you saw them you would not know that they were Miao."[64] The Zhongjia, except for those in Guiyang and Anshun, "are rather docile and good. Recently they would rather put away their weapons. Their cruel customs are returning to pure [ones]. Many can read and write."[65] This new material is significant for the effort it makes to distinguish between Zhongjia who are "docile and good" and would rather put down their weapons and those who are still cruel and to be feared. The differing character of various Zhongjia is associated with the region where they live and how densely populated it is by Zhongjia—vital information for officials governing the region.

The Yanghuang, Dongren, and Hong Miao entries all emphasize relatively recent acculturation. In the case of the Yanghuang, surnames are listed for the first time: Yang, Long, Zhang, Shi, and Ou.[66] Then, in a significant departure from the content of the 1673 and 1692 Guizhou Gazetteers, we are told that their clothing, ornamentation, weddings, and mourning rituals are like those of the Han. In a final sentence the reader is told that of late civilization and culture *(yiguan wenwu)* gradually flourish day by day.[67] The 1741 Dongren entry states in the last sentence that "now they all desire to put away their weapons," a comment not found in the earlier gazetteers.[68] New information added in the last line of the 1741 Hong Miao entry also alters the impression conveyed by

---

two kinds although listed together in one entry. Both the 1560 and 1608 entries on the Kemeng and Guyang Miao explicitly state that the Kemeng and Guyang are two kinds of Miao.

62. *GZTZ* (1673), *juan* 29, *Miao liao* subsection, entry no. 3.

63. *jin yi jin bian er wei han. GZTZ* (1673), *juan* 29, *Miao liao* subsection, entry no. 22.

64. *GZTZ* (1741), 7:18a.

65. Ibid., 7:11a.

66. Ibid., 7:20b.

67. Ibid., 7:21a.

68. Ibid., 7:22a.

the earlier version. We are told that since the suppression *(jiaofu)* they have become docile and obedient *(xunfu).*[69] The 1741 Qing Miao entry, in a similar vein, mentions that they know how to fear the law. These comments on the increasing amenability of various groups to Qing rule and Han customs show that this topic is an issue of concern to the author. The gazetteer reflects a certain degree of success in carrying out policies of assimilation. The fact that much of the 1741 gazetteer was compiled under Ortai's tenure as governor-general of the province may account for this emphasis. As discussed in chapter 4, his policies, unlike those put forward later under the Qianlong emperor, tended to favor assimilation to Han ways.

Li's 1834 *Qian ji* occasionally indicates that certain customs or practices no longer existed as described in *Qian shu*, a 1690 account that is nearly identical to the 1673 Guizhou Gazetteer. For example, the 1834 Qing Miao entry states that whereas *Qian shu* portrayed them as "strong and fierce and liking to fight" *(qiang han hao dou),* now they are "amenable and good" *(shun liang).*[70] Similarly, the Hua Miao entry states that "*Qian shu* records their moon dance in the greatest detail, but now they know to use a go-between."[71] Other statements measuring acculturation, which do not specifically invoke *Qian shu,* also appear with regularity. This emphasis on assimilation coupled with deliberate downplaying of diversity and diminished attention to information gathering seem to be characteristic of the nineteenth century.

Although the 1834 text boasts more entries, compared with the 1741 gazetteer they are much shorter and therefore have a formulaic ring to them. *Qian ji* apparently dates from an era when the Miao albums had ceased to function as administrative documents but were rather objects of interest or curiosity among a more widespread literati population. Late developments in the album genre are described in chapter 7.

From 1560 through 1741 we see an increase in the numbers of categories of *miao man* groups in Guizhou province, and in the geographic area in which investigations were made. Administrative concerns, at issue throughout, are increasingly represented as goals accomplished. As more information is gathered over time there is reason to believe that it has been collected through observation of the groups in question. This

69. Ibid., 7:14a.
70. Li Zongfang, *Qian ji,* 3:2b.
71. Ibid., 3:2a.

conclusion is based on the type of information (observable), and the fact that it comes from locations not previously mentioned in earlier texts. Finally, the 1834 text does not continue the gradual trend established in the development of the earlier texts. Although the number of groups has increased dramatically to eighty-two, the actual amount of information presented has decreased considerably as reflected in the length of individual entries. In addition, relatively few new place names are mentioned.

The development of ethnographic writing in Guizhou province during the course of its colonization from the late sixteenth to the mid–eighteenth century reveals trends not unlike those described by Hodgen and Pratt for the development of early modern ethnography. The kinds of information included, growing evidence of the role of direct observation in collecting data, and the development of a complex system of taxonomy and classification for the groups correspond with developments of the early modern period in Europe. Just as mapping of territory allowed for increased knowledge of and control over physical geography, depicting peoples was a way of knowing, and a means to better controlling, the human geography of areas into which the Qing was expanding.

CHAPTER SIX

Miao Albums: The Emergence of a Distinct Ethnographic Genre

The English term "Miao album" refers to a genre of illustrated manuscripts that describe non-Han groups in various parts of south China, including Guizhou, Yunnan, Guangdong, Hunan, and Taiwan. The genre is defined both by its subject matter and by its form—illustrations and texts that delineate and categorize the different groups that lived within a given province, or occasionally a smaller administrative unit. Each named group receives one entry, consisting of an illustration paired with textual description. The number of entries contained from album to album varies dramatically, even for the same province. While eighty-two is the most frequently recurring number for Guizhou province, albums for Guizhou with as few as seven or as many as one hundred illustrations also exist.[1] With the exception of two albums printed by woodblock that bear no printer's markings, all of those I have examined were executed by hand. Each is thus a unique document and work of art.

The genre began as a kind of administrative document made by, and intended for the use of, officials responsible for administering the peoples represented therein. The content was at least partially based on the ethnographic portions of the local gazetteers described in chapter 5. The thirty illustrations in the 1692 gazetteer closely resemble the composition of some of the illustrations found in certain Miao albums. Yet a substantial amount of variation is seen in texts and illustrations from album to album.

1. Diamond mentions an album with as many as 124 entries; I have not seen it, however. See her "The Miao and Poison," 2.

Besides the gazetteers, additional evidence also suggests that the Miao albums have precursors in illustrated administrative documents that reach back into the Kangxi reign (1662–1722). At least twice the emperor requested drawings of minority groups to be made and submitted to him. In the forty-first year of his reign (1702), the emperor issued an edict in which he recounted what had been recorded in illustrations *(tuyang)* of the Yao of Guangdong that had been submitted to him. His edict reiterated information on population, dwelling areas, and strategic considerations involved in pacifying the Yao. He stated that although their numbers were not great and the land they occupied was not extensive, the mountains where they lived were dangerous and the roads narrow. Although the Yao would not dare to engage in battle by day, by night some of them who were very familiar with the local pathways would approach the Qing army undetected and carry out attacks.[2] The kind of information contained in this edict concerns issues related to the pacification of this particular group by Qing forces. Another submission of illustrations to the Kangxi emperor occurred several years later. The governor of Guizhou province, Chen Shen, after taking up his post in 1707 made detailed inquiries into the circumstances of Guizhou native chiefs and the Miao areas. Following his researches he had illustrations made to which he added explanatory text. Upon their completion he submitted them for imperial inspection.[3]

These official communications relate two cases where annotated illustrations of minority peoples submitted to the Kangxi emperor were made for the specific purpose of gathering information about border and minority peoples that the central government wanted to control. This information on the existence of what seem to have been early album prototypes, taken together with the purposes of the albums as stated in the prefaces discussed below, supports the view that the Miao albums originated as a type of administrative document. Because the vast majority of albums are anonymous and bear no date, sorting out the course of their development is a complex task. Furthermore, since they were unpublished, important recent scholarship on the printing industry in Qing China sheds no light on their manufacture and distribution.[4] But

2. Chuang Chi-fa, "Xie Sui zhigong tu yanjiu," 773.

3. Ibid., 773.

4. On the publishing industry, see Ellen Widmer, "The Huanduzhai of Hangzhou and Suzhou: A Study in Seventeenth-Century Publishing," *Harvard Journal of Asiatic Studies* 56 (June 1996): 77–122; Cynthia J. Brokaw, "Commercial Publishing in Late Imperial China: The Zou and

judging by their gazetteer prototypes we can be fairly sure that the earliest albums emerged sometime between 1692 and 1741.[5]

Physical Characteristics of the Miao Albums

In Chinese the albums are usually generically referred to as *Miao man tu* (Illustrations of Southern Barbarians), or *Bai Miao tu* (Illustrations of the Hundred Miao).[6] The original titles, still legible on the outside of some albums, vary widely in their wording and content.[7] The outer covers of the albums vary in materials, style, and in their degree of ornamentation. Wooden covers may be plain and smooth, or display intricate carvings. Less common brocade covers also encompass a range of quality and design. Some albums in European and American collections have been rebound with Western-style bindings. One two-volume set in the British Library is stored in an ornate box covered with marbled paper that invokes European, rather than Chinese, printing practices. Unfortunately, rebinding can involve some alteration in the original works. Sometimes the texts were discarded. One album, when rebound, was trimmed down from its original size. Not only are the illustrations no longer centered top to bottom on the page, but the bottom half of some of the characters in the text have actually been cut off.[8]

Ma Family Businesses," *Late Imperial China* 17.1 (June 1996): 49–92; and the other articles in the same issue of *Late Imperial China,* which is devoted to "Publishing and the Print Culture in Late Imperial China." See also Dorothy Ko, *Teachers of the Inner Chambers: Women and Culture in Seventeenth-Century China* (Stanford: Stanford University Press, 1994), especially chapter 1, "In the Floating World, Women and Commercial Publishing." In my opinion, her work has relevance for the study of comparative early modernities, although she chooses not to pursue this line of inquiry.

5. Lombard-Salmon came across several references to Ming dynasty works in her research on Guizhou that, judging by their titles, may not have been unlike the Miao albums of the eighteenth and nineteenth centuries. Gao Ru's *Bai chuan shu zhi,* a Ming catalog, mentions the title "Guizhou zhu yi tu," (Illustrations of All the Barbarians of Guizhou). Another catalog, Qian Zeng's (1629–1699) *Du shu min qiu ji,* also mentions a "reedition" of this work that appeared already in 1434, and contained thirty-five images (Lombard-Salmon, *Un exemple d'acculturation chinoise,* 281). Without observing the work itself it is impossible to tell to what degree it approximated the Miao albums of the Qing dynasty. But Jaeger finds that the albums as we know them today could not have been made prior to the Yongzheng period. As he points out, several textual entries consistently record dates of specific events that happened during this reign. In Jaeger's opinion, Miao albums probably began to be produced c. 1740 during the early part of the Qianlong emperor's reign following on the heels of the Yongzheng suppressions. Jaeger "Über chinesische Miaotse-Albums" (1916), 275–76.

6. "Hundred" not in its literal sense, but in the figurative sense of "many"—although a handful of late albums do name precisely one hundred different groups.

7. See appendix for a full list of album titles.

8. Album no. 43.

The album leaves are usually made of paper, but silk or rice paper was sometimes used. It was common for an illustration on paper to be mounted with a silk border rimming the page.

The physical size and proportions of the albums vary almost as much as the number of entries per album. They range from nearly pocket sized—the smallest one I have seen measuring only approximately 7.5 inches high and 5 inches wide—to others more nearly resembling what we might think of as a coffee-table book. Three of the larger albums, for example, are approximately 15 by 15 inches square; 20 inches tall by 12 inches wide; and 16 inches tall by 11 inches wide. The majority of albums fall in between these two extremes.[9]

At least four "albums" are actually scrolls, some of which may have made by remounting album pages for more effective display. Two separate articles by Song Zhaolin, ethnographer at the National Museum of History in Peking, describe two separate sets of four scrolls each, each scroll displaying seven different illustrations.[10] A hanging scroll housed in the Museum of Ethnology in Leipzig, Germany, displays two album leaves and contains a blank space apparently intended for a third. A much larger hanging scroll, housed in the Italian Geographical Society in Rome, has a different appearance. It takes the form of a vertical landscape painting, with ten different minority groups nestled into the scenery. Descriptive text is written on the painting near each group. This work was originally painted as a hanging scroll.[11]

Within the book-like albums, the placement of the illustrations and text in relation to each other is not standardized. Typically the illustrations appear on one page and the text on the facing page. Usually the illustrations will be on the right and the text on the left, but occasionally the arrangement will be inverted, with the text to the right of the illustration. When there is more text than will fit onto the same amount of space occupied by the illustration, several pages may be inserted following each illustration.[12] It is also common for the text to be written directly onto the illustration, just as inscriptions of poems appear on Chinese paintings. In such cases the illustration and text together may occupy one page, the facing page being devoted to an illustration of

---

9. See appendix for measurements of each of the albums.

10. Song Zhaolin, "Qingdai Guizhou minzu de shenghuo huajuan," *Guizhou wenwu* (1987): 33–38; Song Zhaolin, "Qingdai Guizhou shaoshu minzu de fengsu hua," *Wenwu yuekan* (1988): 82–90.

11. Album no. 34.

12. This situation occurs in the 1673 *GZTZ;* in album no. 15; and in album no. 60.

another group, or they may occupy a full two-page spread. Less common is an arrangement whereby a single illustration occupies a two-page spread, and the explanatory text follows on the succeeding two pages.[13] Also somewhat unusual is a situation occurring only in albums printed by woodblock where the illustration and text appear on opposite sides of the same leaf.[14]

Most of the albums bear no indication of the artists' identity—probably because they were unwilling to sign their names to this kind of work. The artists who made the albums did it for the money, not because they considered such work as art. Those who may have had a certain stature as artists could have lost face by taking on this kind of work.[15] Some of the albums may have been done in a workshop employing more than one artist, one painting the figures, another the background scenery. The quality and style of the painting varies from album to album. Some illustrations are quite lively, others more static, some could be characterized as "primitive," others reveal fine brushwork.

Albums were sometimes based on, if not directly copied from, other albums. For example, the poses that individual figures assume are sometimes very similar in two or more different albums, but their placement on the page may differ. In some cases the artist appears to have copied the figures from an earlier album or model, but has taken the initiative in deciding how many figures to include and where to position them on the page. Some albums closely resemble each other, whereas in others the artists seem not to have been constrained by established models.

Additional variable characteristics of the albums include the style of calligraphy, for example, the number of different hands and/or styles that may be included in an album; presence or absence of seals; inclusion of poetry; and incorporation of prefatory material.

BASIS FOR ILLUSTRATIONS

By the eighteenth century texts describing *miao man* peoples were almost certainly based primarily on direct observation. It is more difficult, however, to determine whether the illustrations were painted from direct observation, were based on information in texts, or whether artists instead relied on their imagination in conjunction with the texts. With

---

13. Album no. 50.
14. Album nos. 19 and 49. These are two copies of the same woodblock album.
15. Personal communication, Liu Jin, Curator, Guizhou Provincial Museum, 1991.

few exceptions the albums depict each specific group in a distinct recognizable activity or location. Each visual trope, once established, was repeated with some variation in other albums, thus becoming the standard mode of representation for that group.[16] A good analogy can be found in European artists' renderings of the inhabitants of the New World. Watercolor illustrations painted by John White from direct observation were reproduced many times by other artists and publishers and distributed in the form of copperplate engravings.

The source of inspiration for the illustrations in the Miao albums may not be a question that can be answered definitively, but we do have some information. In the few cases where prefatory material precedes the albums, we can find hints on this topic. The 1786 preface to "Miao man tuce ye,"[17] an album now housed at the Library of Congress, offers the following information about the illustrations and their purpose:

> [W]ith my [own] eyes I have observed the conditions [of the various groups] and from their demeanor can distinguish between them. Their disposition and customs (*fengtu*) are different, therefore according to their type I have drawn this complete [set of] pictures to classify them in one volume. . . . Leafing through and admiring them, one can visualize the people's customs and their uncultivated disposition, each in accord with what is suitable [to them], and the disposition of the ordinary people, pure and peaceful, each following their preferences. Although their clothing is different, and their languages are not mutually intelligible, yet they diligently cultivate the earth and place importance on sericulture. What they devote themselves to prospers, and gradually they are cleansed of vulgar barbarian customs. Human talent daily flourishes, and the people pay tax willingly and associate with each other. It is a moving sight.[18]

It seems important to the author that the reader be able to "visualize" the different peoples. The emphasis, however, is on the prosperity and docility of the peoples portrayed, and more importantly on the effective administration of the region. Verisimilitude in the illustrations is not stressed; the desired effect is rather an overall visual message of the prosperity and law-abiding nature of those represented.

In the preface to "Bai Miao tuyong" (Illustrations and Verses on the

---

16. Two albums I have seen deviate from the standard format in terms of the organization of content. Each entry, rather than describing a specific group, shows a specific activity or location according to which that entry is labeled; for example: buffalo sacrifice, paying taxes, school (*yixue*), etc. These are album nos. 81 and 79. Several of the illustrations in the former album are reproduced in Lombard-Salmon, *Un exemple d'acculturation chinoise*, plates 5, 6, 7, and 9.

17. Album no. 41.

18. Ibid., preface, 1.

100 Miao), an album of relatively late date, housed in the Guizhou Minority Nationalities Research Institute (prefaced 1890), the author offers the following information on the origin of the illustrations:

> In present times, Guizhou's Miao customs are recorded in detail in gazetteers of Guizhou, however [now] there are no illustrations.[19] I am an official of this place. Although daily I see the bustling activity of the Miao people, yet I do not know the differences between the different types of Miao. I have served in office in Guizhou for twenty years, and I am very familiar with Miao customs. (Therefore) I had illustrations made and prepared them to be looked at. At this time I have a colleague by the name of Yu Yushan. He is skilled at painting and drawing, and moreover has traveled in Guizhou over a long period of time. He has encountered with his sight the appearance, clothing, and customs of the Miao, and what they like and are good at, and taken them into his mind. Having them in his mind, he echoes [their image] with his hand [in drawing]. He has also consulted gazetteers and availed himself of notes. Altogether he has painted one hundred [different] types.[20]

Clearly, the artist had some firsthand experience of Guizhou's minorities. Yet he also made use of written materials to supplement his own observations. It is quite unlikely that he actually painted from models or on site. Rather, as in much landscape painting of the period, the artist probably drew from memory and according to his own inner impressions of the scene. The "liberties" taken by the artist in this case would ostensibly be fewer, however, since the goal, as the author describes above, was to show what the groups actually looked like.

In contrast to the familiarity of the artists with their subjects implied by these prefatory remarks, the album illustrations themselves indicate that at least in some instances artists may have based their drawings on texts, or on their own fancy, rather than on direct observation. An examination of various depictions of the Madeng, or stirrup, Longjia selected from different albums illustrates this point.

Judging from the content of the pictures, and as explicitly stated in some of the texts, the name "stirrup" Longjia was probably derived from the distinctive hairstyle of the women. Although in each album a stirrup-shaped hairdo is portrayed on one or more of the figures shown, the variety of hairstyles that can be construed to fit this description is significant. In the albums I have viewed, there are no fewer than five distinctive styles that all could be described as stirrup-shaped but

19. While the 1673 and 1692 Kangxi Guizhou Gazetteers did contain illustrations, later editions were not illustrated.

20. Album no. 15, preface.

otherwise bear very little resemblance to each other. In some cases the hair is gathered into the shape of a hollow ring on top of the head. In others a blue cloth is placed over the head and hangs down to the shoulders on both sides, and a large round bun or hair ornament sits directly on top of the head. Sometimes a kind of closely fitted bonnet covers a bun behind the head and shades the forehead in front, where two stiff sides come together in the shape of an upside-down V. Occasionally the hairdo would seem to require the use of several wooden sticks in order to make a hollow square area above the head. In yet another instance several of these elements are combined. One illustration shows a hairdo where a scarf drapes down from behind the head to the shoulders but also rises above the head coming to a point, under which it leaves a hollow space. The variation in the depictions of the Stirrup Longjia cannot definitively answer the question of whether illustrations were made from direct observation or not; the lack of uniformity might stem from a diverse set of interpretations of the text, but it could also be the product of various artists' interpretations of what they had seen, albeit not at close quarters. One also needs to consider the possibility that the Madeng Longjia may not have had a uniformly observable, characteristic hairstyle.

CONTENT OF THE ALBUMS

What kinds of information do the albums actually convey? This question can be answered in two ways, depending on whether we look at actual information they contain or the more subjective overall impression they create for the viewer. In terms of textual content, the first and most consistently reported information is where a given group resided. From each album entry we learn to which administrative jurisdiction the group described belonged. The second most frequently mentioned categories of description are appearance, including dress and hairstyle, and the nature, or disposition *(xing),* of the people. Appearance is important because for an outsider it is the major identifying feature of any group. As described above in chapter 5, the names of different groups are most often based on a combination of a designation for the broader ethnic group to which a people belonged (Miao, Zhongjia, Gelao, etc.) and some feature of their appearance, often dress or hairstyle. Occasionally a name will be associated with the location in which a group lived, a preferred type of dwelling, a family name, or a particular custom, but appearance was the overall determining factor in the system of categorization. The "nature" of a group was important because administrators needed to know if they should be particularly wary in dealing with certain peoples.

Other topics frequently mentioned in the album texts include courtship and marriage practices, religious practices, death ritual, and seasonal festivals. These topics were of interest to the Confucian literati because of their own strong beliefs in the importance of proper rites. Means of livelihood are also mentioned, and, less frequently, surnames, diet, or other distinctive habits are commented upon. There were pragmatic reasons for wanting to know much of this information. For example, seasonal festivals would bring together large groups of people. It was helpful for officials to know in advance when such gatherings would occur and what to expect.

The contents of the illustrations include sketches from everyday life that often focus on livelihood, courtship practices, religious customs, violent behaviors, or other distinctive characteristics. With few exceptions the albums depict each specific group in a distinct activity or location. Each visual trope, once established, was repeated in later albums, thus becoming the standard mode of representation for that group (see table 1 below). Up to the mid- to late eighteenth century this standardization of images can best be understood as a codification of knowledge about each group rather than as a lack of ongoing interest and observation.

Of the eighty-two non-Han groups that are most commonly identified in the Miao albums of Guizhou province, thirty-four have illustrations depicting in some way the livelihood of the peoples in question (see, for example, plate 7). At a glance a viewer could gain an impression about how the groups fit into the local economy. Activities portrayed include practicing domestic agriculture; gathering medicinal herbs and other plants from mountainous regions; hunting; fishing; spinning, weaving, or dyeing; blacksmithing; and engaging in commerce.

Knowing what a group does for its living conveys a sense of whether its members are productively occupied. Understanding what people do with their days cuts down on speculation about what kinds of subversive activities they might carry out in their spare time. Defining their niche in the local economy makes them appear a productive part of the larger society, with a presumed interest in stability and therefore good reasons to cooperate with local officials. Knowing a group's economic niche also helps one to peg social position.

Courtship and marriage occupied a significant amount of ethnographic attention (see, for example, plate 8). Fifteen of eighty-two illustrations are devoted to some aspect of courtship or marriage ritual in most of the albums. Attention is given to the setting in which young men and women meet each other. Sometimes a seasonal festival serves this purpose. The Bai Zhongjia choose a location each year in the first month of spring where they hold a kind of dance. The men beat on a homemade drum, and the "men and women embrace each others' waists and sport together. Their parents look on and do not forbid it."[21] The Chezhai Miao also choose a flat, open area that they call a moon field *(yue chang)* where men and women come together. The men play stringed instruments and the women sing with a sound that is both clear and beautiful. The Hua Miao also gather in a moon field, where the unmarried men play a reed instrument called a *sheng,* and the women shake bells. They sing and dance throughout the night. The Gouer Longjia erect a spirit pole *(guigan)* in springtime under which men and women dance. The unmarried Kayou sing and dance under the moon in the wilderness. The Xi Miao and Lingjia Miao gather together in a similar way, choosing their own mates. Among the Bazhai Hei Miao, the women put up a house in the wilderness in which unmarried men and women can gather. The Yao Miao have a similar practice. Qingjiang Miao youth hold picnics at which plenty of alcohol is consumed, and then pair off.

In some of these groups the young people simply elope; no betrothal arrangements or wedding ceremonies are mentioned. In other cases betrothal gifts are involved, but sometimes not paid for a full year (Bazhai Hei Miao and Liudong Yiren), or until a child is born from the union (Xiqi Miao). Among others, such as the Gouer, Hua Miao, and Kayou, a matchmaker is employed only after a couple has decided to marry. Delayed-transfer marriage is also noted in several cases. Among the Baizhai Hei Miao, the woman visits the home of her husband for three days after the marriage but then returns to her own home. Kayou, Lingjia, and Xi Miao brides are said not to move into their husbands' households until they have a child.

Other customs revolving around marriage are also mentioned. The Daya Gelao practice an unusual custom in that the front teeth of the bride must be broken off before her marriage. The precaution is undertaken in order to avoid harming the family of her husband.

21. Ruey, *Miao luan (man) tu ce,* 68 (album no. 40).

Among the Caijia, a widow can be reclaimed by her family after the death of her husband in order to avoid being buried along with him. The Liudong Yiren, once having decided on a partner, divine for an auspicious time to marry. During the wedding itself, neighbor women see the bride to her new home, carrying a parasol over her head. Among the Songjia, on the wedding day a representative from the groom's family comes to fetch the bride on his back while the members of her own family pursue the "wife-snatcher," trying to chase him off. All of these practices are vividly illustrated.

Religious rituals, including both worship and divination, define other groups (see, for example, plate 9). The Bai Erzi are depicted in front of an alter where a pig's head, candles, and other offerings have been placed. The Guoquan Gelao place a tiger head made out of flour on their altars; the Manren, an ox or water buffalo's head. The Boren kneel at an outdoor altar displaying sutras and candles. The Luohan Miao are pictured in front of a seated Buddha-figure. Animal sacrifice is also depicted. Mulao are shown sacrificing a chicken; a small straw dragon or boat into which flags of five different colors have been stuck is also part of the scene. The Pingfa Miao are shown preparing to eviscerate a dog. The Hei Shan Miao practice divination with blades of grass, the Hei Jiao Miao with spiral shells. While death ritual is mentioned in many of the texts, the Liu Ezi have the distinction of being the only group in which such activity is illustrated. They are shown washing the bones of their ancestors.

Part of the goal of the albums was to distinguish between non-Han groups that were docile and those that posed a danger. As one preface stated, "Their customs are, of course, distinct and their natures, consequently, can also be distinguished."[22] Typically, albums for Guizhou portray eight out of eighty-two groups as having customs and natures that are violent and threatening (see, for example, plate 10).

The Hei Sheng Miao are shown armed, and they carry off trunks and other goods by torchlight. "Their disposition is cruel. . . . They inquire where wealthy people dwell, and then by the light of torches, armed with weapons, they plunder that place."[23] The Hei Zhongjia are similarly armed with torches, and carry between them a red flag, an axe, a hoe, and a shallow basket. Their target is somewhat more specific,

---

22. Album no. 4.
23. Ruey, *Miao luan (man) tu ce*, 60 (album no. 40).

however. If a debt is not repaid, the lender will seek out the gravesite of a powerful family, dig up the bones of the corpses by night, and leave a flag in front of the tomb with the name of the one who owes him money.[24] By this clever ruse, revenge would be taken on the recalcitrant borrower without the lender having to expend additional resources. Desecration of graves is anathema in Confucian culture, where ancestor worship is part of everyday life.

The Hong Miao, Jiuming Jiuxing Miao, and Zijiang Miao are also portrayed as violent, but here the violence is directed inward at members of the same group. Hong Miao men are typically depicted fighting with one another while the women attempt to hold them back and make peace. The portrayal of the "crafty and fierce" Jiuming Jiuxing Miao is similar, only it is other men, not women, who do the restraining. The Zijiang Miao are especially belligerent. According to some albums texts, they "value life lightly and like to fight. . . . When they obtain the flesh of an enemy they eat it raw."[25]

The Qingjiang Zhongjia and Yangbao Miao posed a threat to Han civilians and officials as well as to each other. The Qingjiang Zhongjia had a reputation for kidnapping solitary travelers, whom they would rob and hold for ransom. In the illustrations the prisoner always appears to be a Han Chinese. The Yangbao Miao would protest visits from official messengers or runners. The village elder (also Miao) would need to use persuasion to make the villagers cooperate.

The Zhushi Gelao are shown drinking a toast as part of a pledge to assist in a vendetta. According to one album,

> When they have enemies they must take revenge. If alone their strength is not sufficient, then they prepare beef and wine in order to treat someone who has the strength to help. For only a catty of meat and a small amount of wine they are willing to sacrifice themselves. They look lightly on life. If someone should come to his death in this way his family would be compensated with a water buffalo.

Although simply looking at the illustrations would not necessarily give one a bad impression of the group—a beautiful flowering vine drapes over a stone wall that forms a backdrop for the pledge, and the liquor is served in dainty cups from a tray—the text is one of the most derogatory. In addition to their propensity for taking revenge, the reader is told that they are dirty and smell bad. Their name probably derives from

24. Ibid., 63.
25. Ibid., 41.

the belief that "in the course of a year do not wash their body or face." They are said to live and sleep together with dogs and pigs, the foul odors being unbearable. Men are shown with, and described as carrying, knives and crossbows.[26]

The remaining groups are identified by relatively neutral or even positive traits. These include hairstyle (Jianding Miao, Jiantou Gelao), costume or dress (Duanqun Miao); musical activities (Bulong Zhongjia, Hei Miao, and Turen); dwelling types (Kemeng Guyang Miao, Louju Miao); or other notable customs. These include dispute resolution (Hei Lou Miao), the practice of couvade (Langci Miao), the existence of female officials (Nüguan), woman with the leisure and education to play chess (Qing Zhongjia), hunting prowess (Jiugu Miao), skill at boating (Dongzai Miao), or extremely tough feet that allow for quick climbing, even while barefoot (Tu Gelao).

Table 4. Commonly Illustrated Tropes in Miao Albums, Listed Alphabetically by Group.[27]

Bafan Miao	Women thresh grain and prepare it in the courtyard of their home. An idle man looks on.
Bai Erzi	Religious ritual involving an altar where a pig's head, candles, and other offerings have been placed. A woman, with infant, looks on from the house. In some albums the scene shows men with two water buffalo instead.
Bai Ezi	Agricultural scene. Men till the earth with hoes, a woman sows seed.
Bai Longjia	Figures are dressed in white. Some are pictured carrying wooden casks on their backs. Others rest with their load beside them.
Bai Luoluo	Several men carry tall baskets on their backs. An alternate trope shows men in a small shelter cooking over a fire with a three-footed cooking pot.
Bai Miao	Pastoral scene including a water buffalo.
Bai Zhongjia	Figures dressed in white play a homemade drum. Others dance.
Bazhai Hei Miao	Courtship scene. Several couples flirt with each other within view of a bamboo dwelling.
Boren	Worship scene. Several figures approach or kneel at an outdoor altar where sutras and candles have been placed.
Bulong Zhongjia	A small group of men play instruments including a drum, horn, and gong.
Caijia	Man approaches over bridge to reclaim widowed family member.

(Continued on next page)

26. Ruey Yih-fu, ed., *Fan miao hua ce* (Taipei: Academia Sinica, Institute of History and Philology, 1973), 22 (album no. 16).

27. Translations of names are found in table 1.

Table 4. *Continued*

Cengzhu Longjia	Woman leads goat, other figures follow carrying baskets on shoulder poles or back.
Chezhai Miao	Courtship scene. Men and women sit together. Men play stringed instruments.
Datou Longjia	Several figures resting from fieldwork, preparing to eat a meal.
Daya Gelao	A group of women surrounds a young bride-to-be, assisting in removing her front teeth.
Dong (cave) Miao	Domestic scene. A woman weaves at a loom. Others assist or look on.
Dong (east) Miao	Pastoral scene including a water buffalo.
Dongjia Miao	Several figures with baskets on their backs walk in natural scenery.
Dongren	Winter scene. Figures return from collecting reeds.
Dongzai Miao	A woman stands near the bank of a river, a man in a boat approaches.
Duanqun Miao	Women appear in very short skirts. Sometimes a hidden man looks on.
Gaopo Miao	Figures dye fabric in a large vat or are shown carrying agricultural implements.
Ge Zhuang	Family scene, including a woman weaving at a loom.
Gouer Longjia	Several figures dance around an upright post resembling a maypole.
Gulin Miao	A woman spins at a wheel. Another winds thread.
Guoquan Gelao	Religious ritual involving a tiger head (made out of flour). Women have distinctive hairstyle.
Hei Luoluo	Men at hunt on horseback.
Hei Miao	Men play reed pipes.
Hei Shan Miao	Armed men are shown practicing divination with blades of grass.
Hei Sheng Miao	Armed men with torches carry off trunks and/or other bundles. The last man looks back over his shoulder, watching for pursuers.
Hei Zhongjia	A group of armed men with torches, a red flag, axe, hoe, and shallow basket walk along by night.
Hei Jiao Miao	A group of men divines using spiral shells.
Hei Lou Miao	Foreground pictures several men with a water buffalo. A pagoda appears in the background.
Hong Gelao	Figures carrying a variety of baskets, one on shoulder poles, walk in the countryside.
Hong Miao	Women restrain men from fighting one another.
Hongzhou Miao	A woman works at a loom outside a house. Others look on.
Hua Gelao	Several figures (men and women) return from the hunt with a deer.
Hua Miao	Men play reed pipe instruments, women shake bells, all dance.
Hulu Miao	Harvesting rice.

(*Continued on next page*)

Table 4. *Continued*

Jianding Miao	Several men are shown. Their hair is done up in the shape of a horn on the top of their heads.
Jiantou Gelao	Preparations for a haircut are underway.
Jiugu Miao	Three men pull on a crossbow; another, in full armor, approaches a tiger with a long spear.
Jiuming Jiuxing Miao	Several men want to fight, others hold them back.
Kayou Zhongjia	Ball toss that is part of courtship ritual.
Kemeng Guyang Miao	A tall ladder leads from a streambed up a steep cliff. Figures appear above and below.
Langci Miao	Seated indoors, a man holds an infant. A woman carrying a tray comes toward them.
Li Minzi	Several figures appear in a pastoral setting with goats and agricultural implements.
Lingjia Miao	Men and women dancing. Long sleeves cover their arms and drape down over their hands.
Liu Ezi	Washing bones of departed family members.
Liudong Yiren	Wedding procession. A female companion holds a parasol over the bride.
Louju Miao	A two-story dwelling is shown with animals occupying the ground floor.
Luohan Miao	Several figures kneel in front of the figure of a seated Buddha. Another figure approaches in a reverent posture.
Madeng Longjia	Family shown on its way to fields with hoes and baskets.
Manren	Religious ritual involving an ox or water buffalo's head set on an altar, and candles.
Mulao	Religious ritual involving the sacrifice of a chicken and a small straw dragon or boat into which flags of five different colors have been stuck.
Nong Miao	Agricultural scene. Rice is transplanted, or figures carry agricultural implements.
Nüguan	Female official with attendants.
Pingfa Miao	Several men prepare to eviscerate a dog.
Pipao Gelao	Several men work at a forge.
Qian Miao	Agricultural scene. Figures with hoes, sometimes shown tilling the earth.
Qing Miao	Pastoral scene including several figures with agricultural tools, and several pigs.
Qing Zhongjia	Two women play Chinese chess.
Qingjiang Miao	Courtship scene in picnic setting. Men wear brocade robes and drink from long horns.
Qingjiang Zhongjia	Several armed men surround a Han prisoner in a cangue whom they kidnapped for ransom.
Ranjia Man	Hunting or fishing scene.
Sheng Miao	Armed men depart for a day's hunting and/or fishing. Women are also depicted.
Shui Gelao	Several figures standing in a stream fish with baskets and nets.

(Continued on next page)

Table 4. *Continued*

Shuijia Miao	Bringing cloth to market.
Songjia	Bride leaving home.
Tu Gelao	Several men in grass cloaks carrying hoes. One man rubs ointment on the bottom of another's foot.
Turen	Male figures with masks and instruments prepare to perform local opera. In some albums women look on from inside a house.
Xi (west) Miao	Men play instruments (reed pipes and a gong), one or more women dance.
Xiqi Miao	Courtship scene. Men play reed pipes. Women carrying small baskets look on.
Yangbao Miao	A Han official and his runner pay a visit to Miao territory. One male Miao restrains another from attacking them.
Yangdong Luohan Miao	One woman works at a loom, another washes her hair in a basin.
Yanghuang Miao	Several figures walking through natural scenery. They carry baskets, and lead a small animal.
Yao Miao	Courting scene. Young women await or entertain suitors.
Yaoren	Figures return from collecting medicinal herbs. They have tall baskets on their backs.
Yaque Miao	Agricultural scene. A male figure holds a hoe. Women carry just-harvested reeds or baskets on their backs.
Yetou Miao	Agricultural scene. Two men plow a paddy. No draft animal is used; instead a man pulls the plow.
Zhushi Gelao	Armed men drink a toast or pledge.
Zijiang Miao	Several male figures engaged in an altercation (of varying degrees of severity depending on the album).

As for the subjective overall impression that leafing through an album would have conveyed, it is defined by a comforting appearance of order, an assurance that affairs in newly acquired territories were under control. The albums conveyed a sense that the frontier was known, that there was an order to the exotic peoples and customs found in these regions, and that by uncovering it the officials could and did know what they needed in order to maintain harmony in the region. Violent groups still did exist, but they were in the process of being better understood, and even domesticated.

Several factors contribute to the aura of orderly security conveyed by the albums. First, in their design and layout the reader is presented with a sense that the albums are somehow comprehensive both in their scope (peoples represented) and in their content. Secondly, the quaint nature of most of the illustrations (see plates 7 through 10) communicates to an observer that there is no cause for alarm or worry; these peoples are for the most part rustic and simple. On a related note, albums would carry the message to upper-level administrators that the situation near the frontier was in control; the Miao were being successfully pacified.[28] Along with this subliminal message conveyed by the illustrations, textual comments about the relative degree of sinicization of a given group are sometimes included, especially in the texts of albums of a later date.

The albums primarily served officialdom, and did so with a dual purpose. On the one hand they recorded information about the jurisdiction under which the various groups resided, gave background on how to identify members of different groups visually, and alerted them to what kinds of traits to watch out for and what kinds of behaviors to expect. Thus they served to make sense of the indigenous population for practical purposes. At the same time they carried the somewhat propagandistic message that the natives, although once restless, were increasingly calm and docile under the wise rule of the imperial government. As documents the albums both enabled and reflected increased control.

AUTHORSHIP

Most of the albums are anonymous, but there are several exceptions where a signed preface has been included. From the prefaces we can learn something about the authors, and about their reasons for making the albums. In each case where an author can be identified he was an official assigned to govern the area in which the peoples described in the albums reside. The reasons given for compiling the albums relate to the duties of governing the area.

An album attributed to a magistrate in Guizhou contains a preface that appears to be based on the 1692 Guizhou Gazetteer. In the opening lines the author remarks that Guizhou is "a desolate frontier region of the southwest. Its settlements are scattered amidst the Man barbarians." Although central government administration of the area had gradually been implemented, beginning already in the Ming Dynasty, "the various tribes have still not been completely changed." The preface proceeds to discuss the variety of groups, and the fact that they differed from one another. The author states that "their customs are, of course, distinct and their natures, consequently, can also be distinguished." He then listed each group by name. His point was that because they are all different even from each other, it is not easy to improve them. The heart of the matter is that "if one does not differentiate between their varieties,

---

28. Albums were sometimes presented to high-ranking officials as gifts. Archibald Colquhoun, *Across Chryse* (London: Sampson Low, Marston Searle and Rivington, 1883), 359–60.

or know their customs, then one has not what it takes to appreciate their circumstances, and govern them." Finally, in the closing lines, he states, "For those who are officials here, there are all these things [to consider]. Therefore we have explained [the different groups] in detail in the illustrations that follow."[29] This album was made for administrators of Guizhou province.

Another preface, the earliest explicitly dated preface I have seen, was composed *(zhuan)* by a certain Chen Zongang, in 1768.[30] The closing lines indicate that it was written in a government office *(guanshu)*, suggesting the official as opposed to private nature of the work. From the preface we learn both about the production of the album and something about the reasons why it was made. The preface states that a certain Mr. Xue, a greatly respected frontier post commander, had collected forty Miao illustrations into one volume in 1766. One year later, in 1767, Xue showed these illustrations to Chen, requesting him to use four styles of calligraphy to pen an explanation following each illustration. Of these forty illustrations, three appear to have been lost. Of the thirty-seven remaining pages there are ten pages each in two styles, nine pages in a third style, and eight pages in a fourth style. Originally there would have been ten samples of each of the four styles. Xue not only had the illustrations painted, but also compiled the text. He tells his readers that he used various local gazetteers as sources.[31]

This album's primary goal, put together as the album was by a Guizhou frontier post commander, was to catalog the differences between the forty groups it depicts. The stated purpose was to bring together and record information in one volume concerning diversity in dress, ornamentation, worship, marriage, and funeral customs. The author's position indicates official interest in collecting this knowledge in order to achieve more effective governance. One of the final sentences of the preface reads: "From the small picture the viewer has the power *(shi)* of

29. Album no. 4. Judging by place names included in the texts, the album dates from some time after 1730. An anonymous album (no. 39), dating from after 1797, contains a preface that is quite similar.

30. For album no. 6, untitled. The album also contains a separate colophon written in a hand unskilled at Chinese characters (almost certainly a non-Chinese) indicating the book was purchased in 1883 at a place called Fulin Tang. I suspect that the album was already of considerable age at that time. A note penciled inside indicates that the Museum of Ethnology in Berlin obtained the album in 1889. In addition to the prose preface, this album also contains two additional prefatory pages containing a total of ten poems, each consisting of twenty-eight characters. Each poem is followed by a two-character description of the activity, custom, or quality described in the poem.

31. Ibid., preface, 1.

(having gone) 10,000 *li.*" The idea of information as empowering was not alien to those who had a hand in constructing the album.

Fang Ting, who held an official post in Guizhou having been sent there as intendant of a circuit in 1784, authored an album entitled "Miao man tuce ye" whose preface, quoted above, is dated 1786.[32] Comparison of its forty-one entries with the thirty-seven remaining entries in the Berlin album shows that only five groups carry an entry in both albums.[33] Thus this album is not a replica of Mr. Xue's earlier efforts. From Fang's remarks we can conclude that the author had some degree of personal contact with the peoples described. He tells his reader that upon arrival at his post he had given his attention to the local people and local matters. He had visited rural villages and witnessed how the different minority groups lived in proximity to one another, not as segregated communities. He observed variety in the costume, languages, and livelihood of the different groups. He also noted that from the time that Guizhou was first opened up in the Ming, Han people began to move to the area, thus catalyzing a gradual process of sinicization. The image he created, in spite of the cultural difference it portrays, is rustic and harmonious.

One additional album, "Diansheng yi xi yi nan yiren tushuo" (Illustrations and Explanations of Yi Barbarians in the South and West of Yunnan Province), has a preface dating from the Qianlong reign (1788). The tone of the preface, written by He Changgeng, is the same as those we have already discussed, but it is somewhat more direct in stating its purpose: the effective government of these groups. The familiar rhetoric of turning toward civilization is evoked once again, but perhaps even more strongly: "All of the Yi barbarians that live intermixed with Han now turn toward civilization *(xianghua),* engage in book learning, and practice the rites."[34] The author goes on to describe their clothing, diet, weddings, and funerals as being like those of the Han people. After painting this picture of harmony and civilization already accomplished, he then turns to discuss frontier areas where "barbarian" *(yi)* customs still prevail. Because these various groups are his responsibility, he must distinguish "even the hateful varieties." He proceeds to tell his reader

32. Album no. 41.

33. Gulin Miao, Luohan Miao, Kemeng Guyang Miao, Liu Ezi, and Pingfa Miao. Zhongjia are listed in the Berlin album, but three kinds of Zhongjia (Hei, Qing, and Bai—black, blue, and white) each have their own entry in the Library of Congress album.

34. Album no. 75, preface.

that he made inquiries about the circumstances of governing the *yi*, and found that the degree of difficulty and ease required to govern them varied depending on the group. As Wolfram Eberhard has already remarked, the preface states explicitly that the album was "not made to satisfy an appetite for sensationalism, but rather so that officials who presided in areas populated by non-Chinese could familiarize themselves with these peoples and their ways, and thus sinicize them more quickly and more effectively."[35]

The authors or compilers of the Miao albums gave pragmatic reasons for their endeavors, but in their work also communicated their assumptions regarding rulership. We see a certain amount of paternalism and cultural chauvinism, but also a degree of openness to other ways and room for diversity within the growing Qing empire. In his 1768 preface, Chen Zongang addressed the differences and similarities between different people *(ren)*. He wrote that in certain respects, such as their customs and the environment in which they live, people differ markedly. However, with regard to the physical body and also the heart/mind—in terms of capacity for thought, intelligence, wisdom, and cleverness—people are the same.[36] While recognizing a fundamental sameness among humankind, the album enumerates the differences between the various non-Han groups in Guizhou. There is not a simple "us" versus "them" dichotomy; nuanced distinctions are made. While those who are under the umbrella of Chinese civilization are different than those being portrayed in the album in terms of the refinement of their customs and rites, even those groups that remain largely outside the sphere of Chinese cultural influence are recognized as different in significant ways from each other. The language used in this preface is the product of a paradigm that viewed the Qing dynasty's vast scope—and the relative peacefulness achieved in face of wide-ranging cultural diversity—as the result of the emperor's goodness and his ability to maintain harmony between heaven and all under heaven.[37]

The 1786 preface to "Miao man tuce ye" also speaks of diversity within the empire and the emperor's civilizing role. The author invokes the Qianlong emperor's greatness and his capacity to stimulate goodness.

---

35. Eberhard "Die Miaotse Alben des Leipziger Völkermuseums," 129.
36. Album no. 6, preface.
37. Of course this is the view the author of the preface reports. It may coincide more nearly with the image the empire wanted to project than to any objective reality.

I reverently consider the son of heaven's virtue and greatness that has been made known. No matter how far away, there is nowhere that it cannot reach. It stimulates others to goodness, and has a refining influence upon them. It extends to the seas and makes [itself] known in vast areas.[38]

The image of imperial goodness acted like a magnet turning people toward the empire.

The prefatory material exhibits faith in the efficacy of good government, good example, and the goodness of the imperial person to influence these diverse peoples. The idiom of unity despite diversity and without overt armed force prevails. During the Qianlong era the albums served both to catalog and, within certain limits, celebrate the diversity incorporated into the Qing empire along its southwestern frontier. The attitudes reflected in the prefaces reveal that the construction of taxonomic systems for classifying the non-Han peoples of the empire coincided with a the period of optimism and expansion that characterized the eighteenth century. Confidence prevailed that expansion was possible and desirable (for all populations concerned), and that a schema could be adopted for the inclusion of many different peoples into the vision of empire. As will be seen in chapter 7, by the time of the Jiaqing emperor's reign this kind of idealism was no longer prevalent.

The ethnographic texts found in the Miao albums were largely based on information recorded in Guizhou's local histories, but the albums are distinct from the gazetteers in several ways. First, they were conceived and designed specifically by and for officials assigned to governing remote areas of the empire. Secondly, whereas the gazetteers dealt with many different facets of the peoples, products, and geography of the province, the albums were concerned uniquely with the customs and manners of non-Han populations. Finally, the hand-painted color illustrations set the albums apart. In their format and their exclusive focus on ethnography, the albums singled out and defined the peoples they portrayed as notably different from the majority population of the Qing, but nonetheless as subjects of the realm. To this extent the albums were instrumental in constructing a sense of a collective "we" that came to constitute the multiethnic nature of the Qing dynasty and later the People's Republic of China.

38. Album no. 41, preface.

CHAPTER SEVEN

The Evolution of a Genre:
Miao Albums as Art and Objects of Study

O ver the course of more than two centuries we see a definite shift in the content and ideology of the rhetoric expressed in the prefaces to the Miao albums. During the Qianlong reign an optimistic assumption that the groups in question would naturally want to turn toward the Qing empire prevailed. The benevolence and goodness radiating from the emperor would naturally bring about this process. During the Jiaqing (1796–1820) and later reigns this type of reasoning ceased to appear. During the mid–nineteenth century suppression was the order of the day; the desire to learn about the nuances between the different groups had given way to general assumptions about the nature of most *miao man* peoples—bellicose and dangerous. This shift in attitude corresponded with a rebellion that rocked Guizhou province from the mid-1850s to the mid-1870s.[1] Furthermore, as the attitude towards non-Han groups gradually shifted to one of general mistrust, the desire to collect up-to-date ethnographic information on each group seems to have waned. Although albums continued to be produced, many of the albums dating after 1797 become repetitive, and the text, as in Li Zongfang's *Qian ji,* is sometimes curtailed, even drastically. Yet, in spite of the changes in attitude towards Guizhou's non-Han populations and in the value of collecting and recording information about them, the Miao album genre continued to thrive even as albums evolved to fill new niches beyond their initial

1. See Robert D. Jenks, *Insurgency and Social Disorder in Guizhou: The "Miao" Rebellion, 1854–1873* (Honolulu: University of Hawaii Press, 1994). On changing attitudes toward the Miao from the Ming to the People's Republic, see Norma Diamond, "Defining the Miao: Ming, Qing, and Contemporary Views," in Harrell, *Cultural Encounters on China's Ethnic Frontiers.*

and basic administrative functions. They served as collectors' items in China and abroad, filling a demand in the nineteenth-century domestic and foreign art markets; they became objects of attention among foreign scholars; and they provided source material for twentieth-century historical ethnographers.

AUTHORSHIP AND ATTITUDES

By the early nineteenth century the album prefaces betray a more militaristic attitude to the control and suppression of *miao man* groups than during the Qianlong emperor's reign. Concrete administrative concerns and justifications for using force to govern such peoples have unmistakably crept into the rhetoric. Forceful control and punishment of egregious misbehavior is the topic of the first few lines of the 1813 preface to Illustrations of Yunnan's Hundred Man Peoples from Three Frontiers.[2] The author's strategy for governing was to make an example by punishing the leaders of those groups who transgressed. He stated that, because the law-abiding are many and the disloyal few, careful and conscientious use of this method could bring them under control.[3] The idiom prevalent under Qianlong of an ordered universe stemming from good government was replaced by that of force coupled with prevention.

By contrast to earlier album prefaces, this one focuses on details of administration, strategies for keeping non-Han groups in line, and military control of the region. It devotes significant space to describing the production of Yunnan's copper mines and problems sometimes created by demands the mines made on local resources, particularly fuel. Distinctions are drawn between different groups based primarily on their behavior (as opposed to dress or customs), and on what kind of measures were necessary to pacify them. Vocabulary describing the nature of various groups is less charitable than that used in the Qianlong-period prefaces. Whereas earlier prefaces broadly characterized the various non-Han groups as "stupid," "simple," "plain," and "rustic,"[4] some of the groups are now characterized much more harshly. The Luohei, for example, are said to be "monkey-like," "fierce," and given to thievery and stealing.[5] The organization of tribes in terms of their leadership—

---

2. Album no. 72, "Yunnan san yi bai man tu." *San yi*, which I translate as "three frontiers," may refer to frontier areas along Yunnan's borders with Tibet, Burma, and Vietnam.

3. Ibid., preface.

4. *zhuo, yu, pu, li.* Album no. 41, preface.

5. *ju, guang, piao duo wei chang.* Album no. 72, preface.

for example, whether they have a chief, and how many households one chief *(mu)* will oversee—is also discussed. Finally, the author recommended military strategies for pacification. His discussion not only proposed exterminating leaders of non-cooperative groups, but extended to a consideration of terrain and weather conditions that made successful campaigns impossible during many months of the year.

The lengthy preface to this album was used as a forum for discussing policy and strategy. The Qianlong-era rhetoric, with its faith in the greatness of the emperor causing peripheral peoples to turn toward the center of their own accord, has been replaced with policy suggestions on other, more militarily based, methods of achieving control over these peoples and lands. This change of attitude did not occur in a vacuum. A bloody uprising had occurred in 1795, one year before the Jiaqing reign began.

Illustrations of Guizhou's Miao ("Guizhou Miaotu"), dated 1843, was commissioned by Meng Shaojun, Venerable Tribal Overseer, Army of the Front. The preface opens with a metaphor involving a play on words. The primary, although apparently originally unrelated, meaning of the term *miao* is "sprout," as in seedling. The word for weeds, *you*, also has the extended meaning of "bad people." The author develops the metaphor of sprouts for good, law-abiding Miao, and weeds for the "bad elements" among them.

> The law-abiding Miao/sprouts have increased to occupy vast acreage; how could it be that among them there are no weeds? As for the weeds, they cause trouble for the Miao/sprouts. Their malady cannot be cured.

He then further develops the metaphor of eradicating or hoeing out the weeds: "What was not planted will have to be hoed up, sticking in a sharp blade to sever its neck." The author compares the good Miao to sprouts that with careful cultivation and nurture yield desirable fruits. The idea is that the Miao/sprouts themselves are desirable, and must be saved from the weeds (by extension "bad elements") that grow among them. Just like in a field or garden, the weeds need to be eradicated so that the desirable plants can prosper.[6] The same harsh attitudes seen in the Jiaqing-era preface discussed above appear again, perhaps even more strongly.

The final portion of the preface to "Guizhou Miaotu" assumes a different tone. We are told that the viewer will see "pure customs filling the universe." The outlook shifts to focus on what is different, but

6. Album no. 18, preface.

"good," about these peoples: "to cause each [kind of Miao] to cherish the new, it is not necessary entirely to reject the old." In other words, difference in itself is not bad. It can even be pleasant to look at if it does not interfere with the objectives of imperial administration. The pictorial content of the album, the author alerts his reader, deals only with the good Miao, not with the bad weeds.

> As for the other [kinds], I have put them aside without speaking of them. What leisure do I have for seeking out the scarred and dirty? Eradicating the violent in order to bring peace to the good, my official duty is that of a Vice Minister of Justice [xiao si kou]. As for the weeds when they have already been cut, how will my Miao/sprouts not be refined?[7]

The consciously stated selection of only groups that are "good" or "pleasant" to look at is a new distinction. It allows the author to express his viewpoints on suppression of undesirable elements among the native populations while at the same time re-employing the standard eighty-two groups of Guizhou province to form the actual content of the album. Framed in this manner the album becomes a testament to what an individual official has accomplished (eradicating the bad and leaving the good) rather than a source of information encompassing both docile and wild groups. The task of determining who is good and who is bad, who needs to be feared and who reformed, had been assumed by the author of the album.

The 1890 preface to Illustrated Verses of the 100 Miao ("Bai Miao tu yong") reveals another shift in attitude towards non-Han populations by holding up the ideology found in the early albums as a kind of glorious past worthy of emulation.[8] It unabashedly romanticizes policies implemented under the Qianlong emperor, and other earlier (perhaps imagined) periods, and calls for their reinstitution. As with earlier albums, the author or compiler was an official in Guizhou province. The preface provides a forum for the author's views on how the nonacculturated peoples of the southwest should be administered, and dealt with. Like the Qianlong prefaces discussed above, this one emphasizes the fundamental sameness of human nature, and the simple need for cultivation of appropriate virtues. The stated goal is not to highlight the difference between Han and Miao, or even between Miao and Miao, but to make the Miao more familiar, more like "us." Its author does not accept the argument that the Miao are by nature unruly or malevolent:

7. Ibid.
8. Album no. 15.

Those who say, "The Miao like to rebel, it has always been so," regard them as having the nature of dogs and goats [feeling that] there is no consistency to when they will rebel and when live peaceably. Those who govern them simply see them as a special category, and wait for them to beg to be transformed. I say it is not so. Heaven has given birth to mankind in order that men might maintain constancy and love virtue. "In a village of ten families, there must be those who are honest." Miao are also [part of] mankind, furthermore they live in China (zhongtu). Would heaven bestow an obstinate and unruly nature on them alone? Men of old said: "If instruction from fathers and elder brothers does not come first, sons and younger brothers will not follow and obey respectfully." If those in superior positions have no method for giving direction and teaching, then those who are inferior become used to fierce and rude customs. Propriety and morals are not prospering because the schools are not flourishing.[9]

In the eyes of this official the duty of the educated person is to pass along "civilization" through instruction and example.

To provide additional support for this viewpoint, the author invokes examples of successful policies followed during the Qianlong and Jiaqing reigns, during which schools were established and Miao thereby transformed into *min*, normal taxpaying subjects, indistinguishable from their (Han) neighbors. He advocates reimplementing such policies.

My native town is Jingzhou in Hunan. I will take an example from [there]. In ancient times [people dwelling there] were called the San Miao, and they were located south of Dongting. They settled there in the time of the emperor Yao. [From those times until now] the generations are distant, and the years long ago. Moreover, one cannot point to exactly the location where they were. From this it is proved that the Miao were transformed into ordinary subjects *(min)*. But this was still a matter of a long time ago, so I will take an example from modern times to explain it.[10]

At this point the author invokes an example from the Jiaqing reign, during which Miao schools were established and a quota determined for Miao Provincial Graduates *(juren)*. He explains that in a certain county some Miao obtained official posts and the native populations of that area were transformed into normal subjects *(qimin)*. The education level of the locality was improved to the point that it regularly produced persons of wide scholarship. The county he describes was given the name Xinhua, "newly transformed," because of the changes it underwent.[11]

The author then turns to the case of his native place. He states that

9. Ibid., preface.
10. Ibid.
11. Ibid.

schools were established there during the reign of the Qianlong emperor and that the Miao had changed their habits and customs from that time.

> [The officials] did not tell the Miao to change their manner of dress, but [the Miao] changed it themselves. [The officials] did not tell them to shave their heads, but [the Miao] shaved their own heads. [The Miao] were no different from the Han people. Is it not [the case that] the influence of poetry and books flooded [over them] and washed them clean of their old impurities? Now supposing that in the Miao regions one were to establish Miao schools *(guan)* and engage teachers to instruct and teach them. [Such a policy] would cause [the Miao] to follow the way of filial piety and respect for elders, and cause [them] to have a sense of propriety [and submissiveness]. [We should] consider setting up as many Miao schools as are needed and set quotas for each exam, passing one or two Miao *juren.* Those who are especially talented can be selected, and chosen to study at an academy and be encouraged to study hard and put forth a vigorous effort. After a long time, today's Miao students [Miao *sheng*] can be the sprouts *(miao)* of another day. Teach them with Miao instruction and the teaching will easily enter [their minds] and they will be moved as if by magic. Then propriety will flourish, and will not their mistakes and bad habits decline? If we distribute education and thus transform the frontier, first the Miao will change into ordinary subjects, and then those people will [even] forget that they were once Miao. Then will there be a need to be assiduously on our guard [against rebellion]?
>
> I hope that those who see these illustrations will not simply take them as playthings *[wan hao zhi ju],* but as a source for teaching and education. [Miao and *min*] will behave and appear the same, and will have the same customs with little difference. Compared with the Tang Illustrations of Meetings with Kings *[wanghui tu],*[12] the recent [illustrations] are more beautiful![13]

In the final lines the idiom of difference also recurs, in relation to carrying out proper and effective administration. It is the official's duty to know about those for whom he has responsibility. Clearly, however, although recognizing differences is important, transformation through acculturation is the desired goal.

The preface dating from the Guangxu reign idealizes the approach taken under the Qianlong emperor. It invokes eighteenth-century efforts at pacification through education and behavioral modeling, and advocates their reinstitution. From the prefaces we see that attitudes toward the non-Han populations of southwest China and methods for dealing with the challenges that they posed to the government changed

---

12. *Wanghui tu* were paintings of the emperors' meetings with foreign kings who came to the court as tributaries. The genre was made famous in the Tang dynasty. See chapter 1.

13. Album no. 15, preface.

over the course of the Qing dynasty along with the nature of the interactions between indigenous populations and the center.

ALBUMS AS ART

The artistic appeal of the albums helps to explain the continuing demand for albums during periods when the peoples they portrayed were
generally feared and denigrated to the extent that they were no longer
objects of serious ethnographic enquiry. It also helps to account for their
transformation from serving a practical purpose related to governance
to becoming objects collected and admired in literati circles. As the albums attained the status of works of art, the paintings became more
elaborate, their content deviating from the standard tropes. Individual
albums took on distinctive styles. As early administrative documents the
albums were usually anonymous, but later they more often carried the
seal of the artist(s) and/or calligrapher(s). Sometimes a number of individuals would combine their talent in one album, as attested by the
variety of hands, and seals, represented in some of the volumes. Albums
circulated far beyond the confines of the province where they were produced. Officials sent to Guizhou liked to be able to take a Miao album
with them as a kind of memento when they left.

By contrast to albums produced during the mid–nineteenth century,
which had become quite formulaic, albums made in the last decade of
the nineteenth or in the first decade of the twentieth century were in-
fused with a new vigor. An examination of three particular albums of
non-Han peoples in Guizhou, each containing more than eighty-two
entries, illustrates late artistic developments in the genre. These are
"Bai Miao tu yong," the preface of which was described at length above;
"Qiannan Miao man tushuo" (Illustrations and Explanations of the
Barbarians of Southern Guizhou);[14] and "Bai Miao tu" (One Hundred
Illustrations of Miao).[15] We know that the first two of these date from
as recently as the Guangxu reign (1875–1907). The third can only be
dated to after 1797, through an analysis of place names in the text, but is
probably significantly later.

Although each of these three albums differs both stylistically and in
content they share several significant traits in common vis-à-vis the earlier albums. In each case the standard format we have seen is extended

14. Album no. 60.
15. Album no. 28.

beyond earlier limits; each album departs in some way from earlier prototypes, asserting an increased degree of individuality. Not only does the
number of entries increase, one enumerating and describing eighty-six
groups, the others each one hundred, but new ground is also broken in
terms of artistic representation or textual variations. The illustrations
tend to be more artistic in that the skill and vision of the artist takes
precedence over the consistent reduplication of the exact subject matter
of earlier albums. In two of the albums the identity of the artist is indicated by the presence of seals and prefaces, showing that the artistry of
the album is no longer secondary to its function and to the subject matter portrayed.

The six-volume album entitled "Bai Miao tu yong" is exceptional for
its high quality and its completeness; it includes prefaces, poetry, a table
of contents, and seals. The date, the sixteenth year of the reign of the
emperor Guangxu (1890), is clearly indicated. The textual portion of
each entry is longer than average, often occupying several pages following the illustration rather than just one. It is one of very few Miao albums painted on silk.[16] The expensive materials used, inclusion of the
artist's seal, and pains taken to include detailed prefaces all indicate that
this album was an object of considerable value.

Each volume is approximately ten and one half inches tall and nine
inches wide. The first volume contains the prefatory material, tables of
contents, and fourteen illustrated entries. The second volume contains
eighteen entries, the third also contains eighteen, the fourth contains
fourteen, the fifth contains sixteen, and the sixth contains twenty, com-
prising a total of one hundred illustrated entries.

The entries all follow the same general format. The illustrations are
labeled with the name of the group written in seal script. On the following page the group name appears again, written in neat *kaishu* (regular
script). The name is followed by a text of several pages divided into three
parts and written in the same neat hand. The first portion of the text
reflects the content of other, earlier albums. The second segment of the
text begins with a line that is raised three spaces. Here quotations from
earlier sources are cited with attributions.[17] Finally, the third portion of
the text contains an explanation of certain important historical or leg-
endary figures such as Zhu Wang and Pan Hu.

Zhu Wang, the Bamboo King, was the founder and leader of the

---

16. Many are mounted with silk borders, but few are actually painted on silk.

17. Quotes are from local histories of Guizhou such as *Qian ji* or *Qian shu.*

Yelang kingdom. This kingdom was contemporaneous with the Han dynasty (206 B.C.–209 A.D.), and its territory covered parts of what are now Guizhou, Yunnan, and Sichuan provinces. Legend has it that one day as a maiden was bathing on the banks of a river a piece of bamboo three segments long entered her body "between her feet," and she was not able to remove it. She heard the voice of a child crying from inside the bamboo, so when she returned home she broke the bamboo open. Inside was a child, whom she brought up. He grew up to be strong and martial. He took the surname Zhu (Bamboo) and became the Marquis of Yelang (Yelang Hou).[18]

According to legend Pan Hu is the dog ancestor of the Yao tribe. A Chinese king promised the hand of his daughter to the one who could defeat his enemies and return the head of their leader to him. Pan Hu, a large black dog, met the challenge, married the daughter, and together they gave birth to the Yao.[19] Inclusion of this kind of folklore indicates that this album emphasized literary sources, possibly even over ethnographic observation in the present.

"Qiannan Miao man tushuo," in two volumes, is the only illustrated album I have seen that was drawn in black ink. The album leaves of fine white rice paper are eleven and one half inches tall and ten inches wide. The two volumes contain a total of eighty-six entries. Eight different prefaces make up altogether twenty pages of introductory material. Four seals appear on the title page. The ink illustrations are finely executed in exquisite detail. Generally speaking there are more figures included and more detail in the background setting than in most of the Miao albums. While many of the tropes of the illustrations are recognizable from earlier albums, the style of the illustrations is different. The black-and-white brush drawings convey the flavor of amateur literati painting rather than that of professional or workshop painting.

The texts, written in several different hands, appear opposite the illustrations. Sometimes the text is signed, as in the case of entries forty-four through forty-nine, which bear the name Yang Delin.[20] Often, although not always, the text has two parts. The first portion, which reads

18. Yuan Ke, ed., *Zhongguo shenhua chuanshuo cidian* (Taipei: Huashe chubanshe, 1987), 101 and 129.

19. For a detailed discussion of the Pan Hu legend and its variants, see Maxime Kaltenmark's entry on Pan Hu in *Dictionnaire des mythologies* (Paris: Flammarion, 1981), 159–61. See also Anne Birrell, *Chinese Mythology: An Introduction* (Baltimore and London: Johns Hopkins University Press, 1993), 118–20, 264.

20. In entry number forty-five the name was written in and subsequently crossed out.

much like some of the earlier albums in its description of the group portrayed, is written one character higher on the page than the second portion. The album was completed during the reign of the Guangxu emperor, sometime during the years between 1880 and 1890. Thus it was contemporaneous with "Bai Miao tu yong."

"Bai Miao tu," or Illustrations of One Hundred Miao, located in the Italian Geographical Society, is unlike the other two albums discussed in this section in that it contains no preface, seals, or indication of authorship. It is similar, however, in its partial departure from earlier conventions. The most notable characteristic of the album is the imagination the artist brings to the task of illustrating the album. Both with regard to the themes of the paintings and the composition, there is a notable freedom from the constraints set by earlier examples of album illustration. This is not to say that earlier albums were necessarily unavailable, or not consulted—occasionally an illustration will contain certain elements present in earlier albums—but the artist, if familiar with earlier models, was clearly not constrained by them. The texts, which consist of both prose and poetry, are written on bright pink paper, a striking departure from the normal white or cream-colored background.

The illustration accompanying the entry on the Luoluo shows two men in mountainous terrain, one of whom is seated on a white horse (plate 11a). The appearance of a horse in the illustration of the Luoluo is standard, but its posture lends new interest in terms of artistic composition. The horse is shown struggling to keep its footing on a downhill incline rather than in a striding posture typical of those in other albums. The two men are also dressed in a way that represents a departure from earlier illustrations of the Luoluo. They both have a large amount of cloth bound around their heads in a conical shape, and each headdress has a long feather pointing directly out the top of it. They wear pink or red scarves billowing around their necks, and the figure standing on the ground holds a large pole to which a generous-sized red flag is attached. Green streamers also trail from the top of the flag staff. Unlike many albums in which the Luoluo are shown returning from a hunting expedition, it is unclear what kind of activity they are or have recently been engaged in; no game is shown. (Compare with plate 11b.)

An illustration departing even further from earlier precedent appears as part of the Hua Miao entry (plate 12a). The Hua Miao are shown dancing and playing the *sheng,* a reed instrument, as is typical, but rather than dancing outside under the moon in pairs or groups, we see a string of ten women lined up behind a man who is playing the instrument.

Each person holds onto the waist of the person before her, with the man leading the train. It looks like a Miao version of the 1950s dance the bunny hop. One other male figure standing off to the side also plays a reed pipe. Although no natural scenery is depicted, the dancers are presumably outdoors because portions of a tall pole with a flag or streamer attached to it also appear on the right and top perimeter of the illustration. (Compare plate 12b.)

Many other departures from the usual tropes are seen as well. I describe a few of them briefly. The illustration of the Bai Luoluo shows preparations for cremating a body (plate 13a). The corpse is placed on a bed of straw or kindling that is just being lit. Although cremation is mentioned in the texts, this is the only instance where I have seen it depicted in an accompanying illustration (compare with a more standard depiction shown in plate 13b). The Six Types (Shui, Yang, Ling, Dong, Yao, Zhuang) are shown playing what looks like a game of blind man's bluff (plate 14). One figure is blindfolded, and the others are gathered around in various semicontorted poses attempting to stay relatively close but at the same time not be caught by the searching arms of the person who is "it." The Jiugu Miao illustration shows three men positioned in the left-hand bottom corner of the illustration aiming various weapons at a monkey in a tree branch appearing in the top right-hand corner (plate 15a). This is an interesting variation on a theme. Most often the Juigu entry shows three men pulling on a crossbow aimed at a tiger (compare plate 15b). The appearance of a monkey or an ape is highly unusual. The Yetou Miao are shown seated around a table on a very large porch structure enjoying a meal. The Jianfa Gelao are shown working an elaborate irrigation mechanism involving a wooden framework with pedals on which three men are seated with their backs facing the viewer. Another man works a mechanical device that controls water flow into further paddies beyond the scope of the illustration (plate 16). The artist of this album appears to have felt free to depart from the well-established visual tropes of earlier albums.

In summary, we can say that those albums for Guizhou province containing more than eighty-two groups were completed significantly later than those with eighty-two or fewer, at least two of them in the late nineteenth century. As a group these three late albums show a greater preoccupation with presentation, whether in artistic terms, material used, quantity of text, or inclusion of seals and prefaces. Evidence that "Bai Miao tu yong" was designed as a collector's item is seen in its labeling of each of the groups with fancy seal script. Not all of the three late albums share the same traits, but each in its own way departs from earlier precedent in quality and design to distinguish itself from those that came before.

"Nicer" albums also circulated through gift-giving.[21] Archibald Colquhoun mentions that he learned of "a series of pictures of the aboriginal people, made by an amateur artist (a gentleman who painted for pleasure)" during a trip through southwest China in the late nineteenth century. The artist was reportedly no longer living, but copies of his work were still extant, for "[o]ne series had been sent to the Viceroy of [Yunnan] as a present" and another was still in the possession of the artist's family.[22]

Further evidence that albums circulated widely in private hands can sometimes be deduced from a note written inside, or from their present ownership. Jotting in the margin of the first entry of an album now housed in Guizhou's Provincial Museum indicates that the work had been in possession of a certain Guo Peiyuan of Henan.[23] Ownership by someone resident in a different province signals that albums spread beyond the immediate regions where they were made. Inside the front cover of the same album we find a collector's seal showing the album had been owned by a family in Anhui with the surname Yao. To come into the possession of families from outside the province, this album may have been either given as a gift, or acquired while someone in the family had been in office in Guizhou, or purchased on private travels as a collector's item.

A fair number of albums may still remain in private hands. In 1991 while attending an exhibit at the Guizhou Provincial Museum, where several album leaves were on display, I overheard a Chinese visitor to the museum explain to his companion that his family had such an album in its possession. I came across another album in the private hands of a Guizhou antique dealer (it was not for sale). It is not possible to calculate how many albums sit "undiscovered" on dusty bookshelves or in locked drawers.

By the nineteenth century the Miao album had attracted interest not only among Chinese literati, but among expatriates as well. In the late

---

21. For more on the importance of the role of gifts in the system of patronage in eighteenth-century China, see Susan Mann Jones and Philip A. Kuhn, "Dynastic Decline and the Roots of Rebellion," in *The Cambridge History of China*, vol. 10, pt. 1, 113–16.

22. Colquhoun, *Across Chryse*, 359–60.

23. Album no. 55.

nineteenth and early twentieth centuries the Miao albums found a popular niche in the foreign art market. Europeans and Americans began to learn about and collect albums when the colorful and attractive works were abundantly available in used book shops in Peking and Shanghai during the nineteenth century.[24] Foreign interest may have even played a part in resuscitating a cottage industry in the manufacture of Miao albums during the nineteenth century. The high number of relatively poor-quality albums that are probably late copies can be explained in this way. It may also account for the small number of albums without texts. To an audience that could not read Chinese, the art work would have held more interest than the texts. Thus certain late copies may never have been completed by the addition of text, or the accompanying texts may have been lost or discarded at some later point, possibly during the process of rebinding. The considerable number of albums that are housed in museums and libraries in Europe and the United States also attest to expatriate interest in the albums during the late nineteenth century.

An undated note found inside an album now located in the Musée Guimet allows us to trace a specific example of how the albums became known by and transmitted to the West.[25] Written in French on an empty envelope that bears the printed characters *Eguo bing ying* (Russian military encampment), it reads:

> I had the intention of keeping this album for myself thinking that it was not of great interest—leafing through the second volume of Bushell I see that Miaozi are mentioned. There are two images at the end of the volume. I knew that they are a semi-wild people, who are tributaries of China but who were never willing to submit.—I send it to you. I bought the loose sheets in Shanxi and had them mounted in Peking. It cost 50 francs altogether.[26]

We do not know who the note was intended for or who wrote it, but it does tell us that albums were available for sale not only in Peking and Shanghai, but in other parts of China as well. The purchase of this album and the fact that it was sent back to France reveal European interest in the genre. The note may have been written by someone on a buying trip for a museum. Such a scenario would explain the logic of initially intending to keep the album but after seeing photographs of

24. J. Edkins, "The Miau Tsi Tribes: Their History" (parts 1 and 2), *The Chinese Recorder and Missionary Journal* 2 (July and August 1870): 74.

25. Album no. 65.

26. "Bushell" refers to Stephen W. Bushell, *Chinese Art*, 2 vols. (London, 1909).

something similar in a volume of Chinese art then deciding to forward the album with an indication of the cost.

The foreign art market's interest in the albums may be the reason for an interesting if tenuous connection between a famous Jesuit artist and the Miao Albums of Guizhou Province. At least three Miao Albums contain forged seals and/or signatures of Giuseppe Castiglione—known in Chinese as Lang Shining—a European painter employed by the Qing court during the eighteenth century.[27] Forged Castiglione seals were not uncommon. According to one source:

> Albums of engravings and drawings representing Chinese figures—mandarins, soldiers, artisans, courtesans, etc., or else foreigners of various nationalities—bearing the signature of Lang [Shining] sometimes come up for public sale both in London and Paris.[28]

In the mind of the forger(s), some link must have existed between the Jesuit artist and the style of representation found in the albums and these paintings of a similar genre to consider it worthwhile to attempt such a subterfuge.

An incident recounted by John Barrow that took place in the Yuan Ming Yuan Summer Palace provides another example of this association between Castiglione and ethnographic paintings:

> The eunuch took me into a corner of the apartment, opened a coffer resting on a support and, with a meaningful glance, told me he was about to show me something that would astonish me. Then he took out of the coffer several large volumes filled with figures drawn in a superior manner and painted with watercolours. These figures depicted the various trades and occupations of the Chinese. They looked as if they were glued to the paper, since there was neither shadow, nor projection, nor distance to give relief. The page opposite each figure contained an explanation of it in Chinese and Manchu characters. When I had leafed right through one volume of these figures I saw on the last page the name of Castiglione, which gave me the key to the mystery.[29]

Whether the artist was in fact Castiglione is highly doubtful, but the association between him and this type of painting is intriguing.

---

27. They are "Fan Miao hua ce" (album no. 16); an album housed in the Victoria and Albert Museum in London; and several album sheets housed in the Peabody Essex Museum.

28. Cecile and Michel Beurdeley, *Giuseppe Castiglione: A Jesuit Painter at the Court of the Chinese Emperors* (Rutland, Vt., and Tokyo: Charles E. Tuttle Company, 1971), 116.

29. Quoted in ibid., 116.

NINETEENTH-CENTURY FOREIGN SCHOLARSHIP

Foreign scholarly interest in the album genre began as soon as the albums appeared on the market in the early to middle part of the nineteenth century and has continued, with some brief gaps, into the present. A significant number of articles about and translations of Miao albums have appeared over the course of the last one hundred and fifty years. Generally speaking, nineteenth-century literature on the Miao albums tends to be written by European and North American expatriate authors about individual albums they encountered while in China. The focus of these articles is primarily on communicating the ethnographic content of the albums. It is ironic that foreign interest in the ethnographic material the albums contained took hold at the same time that active compilation of such material within China had begun to wane.

The earliest article about the Miao albums appeared already in 1837.[30] The author, Carl Friedrich Neumann, was a German student of Armenian and Chinese languages and literatures. Sometime after 1825 he had traveled to China to purchase Chinese books. During his stay he acquired approximately 12,000 volumes, most of which he gave to the Bavarian government after his return home. In 1833 Neumann was named Professor of Armenian and Chinese Languages, Lands, and Folklore at the University of Munich.[31]

Neumann's article provides a detailed description and complete translation of an anonymous album containing seventy-nine entries, which a Mr. Clarke, of Canton, had shared with him.[32] Neumann hesitates to draw any conclusions about the age of the album, nor does he address its probable uses. The album to which Neumann had access apparently represented the various groups in a highly unfavorable light; he comments extensively on a tendency for the Chinese to look down on non-Chinese. From the context we can deduce that Neumann's own experience as an expatriate living in Canton helped to shape his opinions on this topic.

30. Carl Friedrich Neumann, "Die Urbevölkerung einiger Provinzen der chinesischen Reiches," *Asiatische Studien* (Leipzig: J. A. Barth, 1837), 35–120.

31. *Allgemeine deutsche Biographie* (1886; reprint, Berlin: Duncker & Humblot, 1970). Entry by Julius Jolly, 529–30.

32. Neumann does not state Clarke's first name. A Mr. George W. Clarke of the China Inland Mission also published a translation of an album that appears as an appendix to Archibald Colquhoun's *Across Chryse.* George W. Clarke did not reach China until the 1870s.

Samuel Wells Williams authored an article that appeared in the *Chinese Repository* in 1845 concerning a Miao album. Williams, a printer by trade, had traveled to China in 1833 under the sponsorship of the American Board of Commissioners for Foreign Missions. He settled in Canton, where he printed and later edited the *Chinese Repository*, an English-language journal about China.[33] While abroad Williams developed a scholarly interest in China that he later used both in government service and in academia. He learned both Chinese and Japanese, and from 1855 to 1876 he worked as secretary and interpreter at the United States Legation in China, where he also served occasionally as chargé d'affaires.[34] He returned permanently to the States in 1876, where he took up the newly created position of Chair of Chinese Language and Literature at Yale. As the title of his article, "Notices of the Miao Tsz; or, Aboriginal Tribes Inhabiting Various Highlands in the Southern and Western Provinces of China Proper," indicates, Williams's focus is not on the album itself, but on its content. He provides a partial translation (forty-one entries), adding only minimal commentary. Williams does not indicate how he came to know of the album, who may have made it, or what it was called. Neither does he posit a possible date.[35]

Although Williams does not comment on authorship, we do learn that the album contained a preface. Because prefaces to the albums are somewhat rare, I quote Williams's translation of this preface in full:

> Whenever I have extended my rambles to other provinces, and noticed remarkable views or objects, I have always taken notes and sketches of them, not that I supposed these could be called fine or beautiful, but because they gratified my own feelings. Still, I think that among all these views and natural objects,—the flowers, birds, animals, &c., there were some singular and rare forms, which may be called curious. Moreover, having seen the people in [Guizhou] province, scattered in various districts and places,—both those whose customs are unlike, and also the different customs in the same tribes, having utensils of strange shapes and uses, not discriminating in their food between that which was ripe and the raw, having dispositions sometimes gentle and at other times violent,—having seen their agriculture and manufactures,—having noticed that the men played and the women sung, or the men sung and the women danced; also having viewed their hunting deer and

33. *The National Cyclopaedia of American Biography*, (University Microfilms, 1967; copyright 1891 by James T. White & Co.), 422; and *The Biographical Dictionary of America* (Boston: American Biographical Society, 1906), s.v. "Williams, Samuel Wells."

34. He also acted as Commodore Perry's interpreter in Japan (1853–1854).

35. Samuel Wells Williams, "Notices of the Miao Tsz, or, Aboriginal Tribes Inhabiting Various Highlands in the Southern and Western Provinces of China Proper," *Chinese Repository* 14.3 (1845): 105–15.

trapping rabbits, which are the products of the hills, and their spearing fish and netting crabs, the treasures of the waters, their manner of cutting out caves in the hills for residences, and of framing lofts from bamboos in trees for lodgments, all of which usages were unique and diverse:—these I thought [sic] were still more remarkable. Then I perceived that there are both common and rare things in the world, and races unlike common people; I therefore sketched their forms on one page, and gave the description on the opposite, in order to gratify my own feelings and those of others who wished to see these things. The following are some of these descriptions.[36]

The preface is unusual in that it does not mention any official position the author might have filled that would have accounted for his interest in Guizhou's minorities. It is not clear, however, whether Williams translated the entire preface or just a portion of it.

In 1859 a translation of an album containing eighty-two entries appeared in the *Journal of the North China Branch of the Royal Asiatic Society*.[37] The author, Elijah Coleman Bridgman (1801–1861), had arrived in Canton in 1830 under the auspices of the American Board of Commissioners for Foreign Missions, the same group that had sponsored Williams. He worked with Williams as editor of the *Chinese Repository* until 1847. Bridgman was a pastor by profession, but was involved in many other pursuits as well. He assisted in the founding of the Medical Missionary Society in China, edited the journal in which the translation under discussion appeared, occasionally provided service to the U.S. government in China as translator and adviser, and devoted the last fifteen years of his life to working on a Chinese translation of the Bible.[38]

Close examination reveals that the album Bridgman describes and translates is the same album that Williams had written about sixteen years earlier. However, Bridgman's is a full rather than partial translation. His assertion that the descriptive passages "were written many years ago by a native scholar who had travelled in the province of [Guizhou]" and that "sketches and pictures were both made from personal observation" may be based on his understanding of information contained in the album's preface. Bridgman also remarks that the "work is a curious and rare one, and the only complete copy, which I have ever seen of it, is now in England and owned by Dr. William Lockhart."[39] Presumably Bridgman's translation was based on this

36. Ibid., 107.
37. E. C. Bridgman, "Sketches of the Miao-tsze," *Journal of the North China Branch of the Royal Asiatic Society* 1.2 (1859): 257–86.
38. *Dictionary of American Biography*, vol. 3 (New York: Charles Scribner's Sons, 1929), 36.
39. Bridgman, "Sketches of the Miao-tsze," 258.

complete copy, although he had learned of the existence of additional, if partial, albums.

Dr. William Lockhart (1811–1896) had served as a medical missionary in China with the London Missionary Society for over twenty years, beginning in 1839. In this capacity he undoubtedly made the acquaintance of both Williams and Bridgman. In 1861 Lockhart authored an article entitled "On the Miaotsze or Aborigines of China" for the *Transactions of the Ethnological Society of London.*[40] His article is concerned broadly with the geography and peoples inhabiting the interior of China. He alludes to an album of his own and also to the existence of others.[41] His primary interest lay in the ethnographic information such albums contain, rather than in posing questions about the development of the genre.

In 1870 Joseph Edkins published a two-part article in the *Chinese Recorder* that makes reference to Miao albums. Edkins, like the other American authors described above, had come to China as a missionary. He arrived in 1848 and served under the London Missionary Society for the next thirty-two years. In 1880 he resigned from the Society and took a post as translator for the Imperial Customs Service.[42]

Edkins's article focuses on the history, customs, rebellions, writing, and languages of the non-Han groups of Guizhou. Miao albums are specifically mentioned in part two of the article, and form the basis for the information the author relates on customs. The article is significant because it contains the first reference to Miao albums as constituting a larger genre. He observes that

> [t]he customs and mode of life of the [Miaozi] are by the Chinese regarded as very curious and amusing. Otherwise, coloured drawings illustrative of their customs and occupations would not be so numerous as they are. Books of these illustrations are common both at Shanghai and in Peking."[43]

Edkins's remark indicates that the albums became popularized in the late nineteenth century, and that they were widely available in urban centers far removed from the places and peoples the albums represented.

G. M. H. Playfair, who served as British Consul in China during the last three decades of the nineteenth century, also wrote an article on the

40. William Lockhart, "On the Miaotsze, or Aborigines of China," *Transactions of the Ethnological Society of London,* n.s., 1 (1861): 177–85.

41. Ibid., 182.

42. Pat Barr, *To China with Love: The Lives and Times of Protestant Missionaries in China, 1860–1900* (New York: Doubleday and Co., 1973), 82.

43. Edkins, "The Miau Tsi," 74.

albums. "The Miaotzu of Kweichou and Yunnan from Chinese Descriptions"[44] describes forty-two different ethnic groups found in Guizhou and Yunnan provinces. The information in the article is based on three separate albums, two on Guizhou and one on Yunnan, all of which had been purchased in Peking (presumably by the author). The entries on different groups are based on a compilation of information gained from the three albums to which he had access.

George Clarke's "Translation of a Manuscript Account of the Kwei-chau Miao-tzu" appears in an appendix to Archibald Colquhoun's *Across Chryse*. Clarke, who joined the China Inland Mission in 1876, made several trips to Guizhou later in the same decade. His translation is notable because of the substantial length of many of the eighty-two entries. Although Clarke does not include any introductory passage with his translation, the subtitle of his work, "Written After the Subjugation of the Miao-tzu, about 1730," reveals his opinion on the date of the work. He does not discuss the basis for this judgment.[45]

Foreign interest in ethnographic works about China was on the rise during a period when European and American powers were making inroads into China. The presence of Western missionaries, businessmen, and government personnel in China corresponded with a period when Western powers were eager to learn more about China, and the Qing was hard-pressed to maintain its authority against both foreign and internal threats (most notably the Opium War and the Taiping Rebellion). Thus, although somewhat ironic, it is not surprising that foreign interest in the albums as ethnographic documents surged even as domestically they were valued more as a curiosity.

TWENTIETH-CENTURY SCHOLARLY INTEREST

A substantial shift in the type of interest shown in the Miao albums dates from the early part of the twentieth century. From this time scholars (both Western and Chinese), often writing about albums housed in European collections, began to write about the albums as constituting a specific genre. Rather than straightforward translations or descriptions of the content taken at face value, much more effort was applied to discussion of the origins and history of the albums, and an evaluation of

44. G. M. H. Playfair, *The China Review* 5.2 (1876–1877): 92–108.

45. George W. Clarke, "Translation of a Manuscript Account of the Kwei-chau Miao-tzu," in Colquhoun, *Across Chryse*, 363–94.

the ethnographic content. We begin to see the development of a critical interest in the sources, manufacture, dating, and purposes of the albums.

F. Jaeger's two-part article on the Miao albums is primarily concerned with precedents to, and with the dating of, the albums.[46] In his opinion the earliest possible date for their appearance would have been during the Yongzheng period (1723–1735),[47] when Chinese control was being imposed on these groups. Jaeger relates the usefulness of the albums to the goal of Chinese administration of newly conquered territory, and notes this as a possible reason for their continued production throughout the nineteenth century. He also discusses the relationship of the Miao albums to the Qing Imperial Illustrations of Tributaries.

In 1930 Chungshee H. Liu, a research student of the Oxford School of Anthropology, made an annotated translation based on seven albums (five of which were illustrated) under the title "The Miao Tribes." He says he was able to discover the author, whom he calls Wang Jin; however, the source of his discovery and indications for its accuracy are not spelled out. He seems to assume that all of the albums were authored by one person, which we now know not to be the case.[48]

Liu Xian, a Chinese scholar affiliated with Shandong University, was the author of the earliest twentieth-century Chinese language publication showing an interest in the Miao albums.[49] His 1933 article enumerates all of the albums of which he learned and their locations (in both China and Europe), and also gives references to previous scholarship on the albums. Like Jaeger, he too explores the relationship of the Miao albums to the Qing Imperial Illustrations of Tributaries. Finally he lists the names of one hundred and twenty-two different groups that he compiled in the course of his study.

Chiu Chang-kong, a student of Professor Thilenius, then director of the Museum of Ethnology in Hamburg, Germany, produced the next piece of literature we have on the Miao albums. "Die Kultur der Miaotse nach älteren chinesischen Quellen" consists of a full translation into German of the album donated in 1890 to a library in Gotha, Germany, by Friedrich Hirth.[50] In addition to the translation, Chiu includes an introductory article on the history and sources of the Miao albums. He gives a fuller account of possible Chinese textual sources than the earlier

46. Jaeger, "Über Chinesische chinesische Miaotse-Albums" (1916 and 1917).
47. Ibid. (1916), 275.
48. Chungshee H. Liu, "The Miao Tribes," manuscript, 1930.
49. Liu Xian, "Miao tu gai lüe," *Guoli shandong daxue kexue congkan* 2 (1933): 25–37.
50. The library is now called the Forschungs- und Landesbibliothek Gotha.

literature on the Miao albums. The article also contains reproductions of each of the illustrations in the album.[51]

Wolfram Eberhard, a German sinologist who taught and published in the United States, authored a 1941 article on the two Miao albums housed in the Museum of Ethnology in Leipzig.[52] The article first addresses the museum's Guizhou album in light of other similar albums (then housed in Berlin, Gotha, Hamburg, and appearing in published sources). Eberhard analyzes each of the groups within these albums, attempting to show that, although the albums between them reveal a total of ninety-nine different names, these names actually can be equated or reduced to a standard set of eighty-two groups. He then graphs each of the names against a series of cultural characteristics, so that at a glance the reader can determine which groups practiced what rites or customs, for example, marriage after giving birth to a first child, hunting, batik, slash-and-burn agriculture, etc.

A third table graphs the names of various different groups named in an album for Yunnan against the various categories of ethnographic information given about each group in the album. Thus again we can at a glance determine which groups are involved in agriculture, which groups have women who are involved in weaving, which live in multi-storied houses, which are sinicized, etc. Eberhard's article also shows the geographical distribution of the various groups, according to both the Guizhou and Yunnan albums.

Eberhard is of the opinion that the Miao albums served a functional purpose. To support this statement he quotes the preface of an album for Yunnan that states the author did not have the album made in order to satisfy his own taste for the sensational, but rather so that officials responsible for non-Chinese areas could learn about these peoples in order to be able to transform them better and more quickly.[53]

Another German sinologist, Herbert Bräutigam, in a 1963 article, "Über Miao-Alben," made a study of twenty-five Chinese ethnographic texts including both Miao albums and related but unillustrated manuscripts. He divided the albums he had seen into four main types depending on characteristics of their illustrations (examples of which are reproduced at the end of his article). He also dates albums according to whether they were made before or after 1797 judging by place names.

51. Chiu Chang-kong, "Die Kultur der Miao-tse nach älteren chinesischen Quellen," *Mitteilungen aus dem Museum für Völkerkunde in Hamburg* 18 (1937): 5–32.

52. Eberhard, "Die Miaotse Alben des Leipziger Völkermuseums."

53. Ibid., 129.

Bräutigam was especially interested in the economy, livelihood, and material culture of the groups. He noted that the earlier *Qian shu* (1608), which resembles the album texts in some respects, contains little if any observation on the livelihood of the groups and on economic and social relationships. He finds that later the government used such information in order to bring the Miao into its own larger sphere of cultural and economic influence.[54] He also notes a concurrence between the locations of uprisings in the eighteenth and nineteenth centuries and the locations about which social and economic information is included. From this he concludes that the officials of southwest Guizhou made it a point to learn more about the way of life and domestic economy of the area in order to incorporate these peoples. In his view this was the purpose, or one of the purposes, of the Miao albums.

In 1973, under the direction of Ruey Yih-fu, the Institute of History and Philology at Academic Sinica in Taiwan reprinted two albums of Guizhou province in their entirety. The introduction, identical in both volumes, describes the eleven albums in the Institute's collection, contains bibliographic information on previous literature, and gives the location of additional albums in other collections. Ruey also discusses possible pictorial precedents to the Miao albums and the linguistic classifications into which Guizhou's non-Han groups, as described in the albums, fell.

A 1987 article appearing in the journal *East and West* concerns a collection of Miao albums housed in the Italian Geographical Society in Rome.[55] The author, Giuliano Bertuccioli, professor of Oriental Studies, University of Rome, introduces the fifteen albums housed by the Society, explores possible Chinese sources, gives the most complete review of the Western literature on the albums that I have seen, and translates one of the albums (inventory no. 57c) in full with detailed annotations. In addition, the article reproduces the same album in its entirety, as well as including copies of sample pages from four additional albums in the Society's collection.

In the People's Republic of China interest in the albums has also revived. Song Zhaolin, ethnographer at the National Museum of History, has authored two articles about albums in the museum's collection, as

---

54. Bräutigam, "Über Miao-Alben," *Mitteilungen des Instituts für Orientforschung* 9 (1963): 297.
55. Giuliano Bertuccioli, "Chinese Books from the Library of the Italian Geographical Society in Rome Illustrating the Lives of Ethnic Minorities in Southwest China," *East and West* 37 (1987): 399–438.

discussed in chapter 6 above.[56] In the 1987 article he describes the contents of the illustrations of four different scrolls each containing seven illustrations of Guizhou's non-Han groups, explores the historical circumstances under which they may have been made, and remarks on the living conditions of the groups described. In the 1988 article he transcribes the texts and reproduces several illustrations from the same set of scrolls. In addition he describes their historical background and their value for scholarly inquiry.

A 1991 English-language article by Vladimír Liščák introduces his dissertation (written in Czech), which is comprised largely of a close ethnographic analysis of a Miao album housed in the Museum of Asian, African, and American Cultures in Prague.[57]

Finally, the most recent publication on the albums was produced by the Institute of History and Philology of Academia Sinica in Taiwan. The booklet reproduces three albums in the collection, one for Taiwan, one for Hainan Island, and one for Yunnan.[58] In addition, the booklet introduces three albums for Taiwan in other collections: "Fanshe caifeng tukao," housed in the National Central Library in Taipei, which purportedly dates from 1745; "Taiwan neishan fandi fengsu tu," housed at the National Palace Museum; and "Zhuluo xianzhi fansu tu," reprinted in *Taiwan wenxian congkan*, no. 141.[59] These albums on Taiwan's indigenous peoples are organized around the activities they engaged in rather than by the ethnicity of different groups.

We saw earlier that European and American Missionaries and scholars who worked in China during the nineteenth century took an interest in the albums because they were eager to learn more about the peoples portrayed. In the twentieth century, sinologists resident outside of China learned of the albums. In contrast to the earlier literature, these authors tended to focus on questions about the development of the Miao album genre itself. More recently Chinese scholars both on Taiwan and in the

56. Song Zhaolin, "Qingdai Guizhou minzu de shenghuo huajuan" (A Qing Scroll on the Life of Guizhou's Nationalities), 33–38, and "Qingdai Guizhou shaoshu minzu de fengsu hua" (Qing Illustrations of the Customs of Guizhou's Miniority Nationalities), 82–90.

57. Vladimír Liščák, "'Miao Albums': Their Importance and Study," *Český Lid* 78.2 (1991): 96–101; see also "'Miaoská alba' jako etnografický pramen" ('Miao Albums' Viewed as an Ethnographical Source"), vols. 1–2, Praha, UEF ČSAV, 1990.

58. "Taifan tushuo," "Liren fengsu tu," and "Dian yi tushuo" (album nos. 82, 80, and 69, respectively).

59. Song Guangyu, ed., *Huanan bianjiang minzu tulu* (Taipei: The National Central Library and the Institute of History and Philology, Academia Sinica, 1992), 31–32.

People's Republic have reprinted or reproduced portions of albums, thus enabling a broader audience to learn of their existence and examine their contents. How they are used and what kinds of questions are asked of the material varies for different audiences.

Various opinions have been put forward on the reliability of the Miao albums as source material for the historical study of ethnic groups of Guizhou province. Eberhard based his article on the premise that the ethnographic information they contained was accurate. Both Beauclair and Lombard-Salmon, on the other hand, have exercised more skepticism in their approach to the Miao Albums. But, as we have seen above, in chapter 5, examining local histories shows that much of the information the albums contained can be independently confirmed by contemporaneous sources which were regularly revised and updated. Yet, as is true today, what is seen and subsequently recorded is refracted through the lens of the ethnographer's own experience and interests.

Initially the albums had a significant functional purpose in relation to governing, and accuracy was a high priority. Some album prefaces explicitly state that they were prepared for official use. However, as the albums became more popularized and were aimed at collectors rather than at administrators, concerns with accuracy in representation may have given way to other interests, such as producing an eye-catching work of art. Although the specific standards for accuracy may have shifted over time along with the contexts and purposes for which the albums were made, throughout the course of its development the Miao album genre provides a record of continuing and multifaceted interest among Chinese officials and a variety of scholars in non-Han peoples.

CONCLUSION

T he Miao albums, Qing Imperial Illustrations of Tributaries, and Kangxi atlases are all examples of early modern visual representations of the Qing empire that have contributed to the definition of the modern Chinese nation-state today. Like maps with different purposes and varying scales, each represented distinctive aspects of the Qing empire in specific ways. The atlases are relatively small-scale geographical maps depicting the territory of the empire, and what lay beyond, whether frontier areas or bordering countries. Their purpose was to enable the select group of people who were allowed to view and to use them as accurate a grasp of the geography of the empire and its frontiers as possible.

The Illustrations of Tributaries, by contrast, represented the people who paid, or could be construed to pay, tribute to the Qing. As such the document provided a symbolic map of those peoples who comprised, or at least recognized, the Qing empire. As a rejuvenated form of an older genre *(zhigong tu)*, the Illustrations of Tributaries also played into the well-established Confucian schema of the emperor (whether of Han or non-Han origins) as the symbolic and virtuous center toward which peoples would naturally gravitate, paying homage and offering obedience. In the section on Guizhou province, individual poses and implements shown in the Illustrations of Tributaries are sometimes similar to those depicted for the same group in a Miao album.[1] Unlike in the Miao albums, however, only one male and one female representative of each group are pictured. Natural scenery does not appear in the background,

---

1. Cf. Jaeger, "Über chinesische Miaotse-Albums" (1916), 276–78.

although agricultural implements, looms, musical instruments, or weapons are sometimes included. (See plates 2, 3, and 11.) The Illustrations of Tributaries portray a more uniform and orderly picture than the albums, which like a large-scale map contain more detail. These differences point to subtle but significant distinctions with regard to the goals and audiences of the two genres.

Comparison of the Hua Miao, Hong Miao, and Zijiang Miao entries from the Illustrations of Tributaries and the 1741 Guizhou Gazetteer, on which many of the album texts were based, shows that while the texts of the two works contain some similarities, they diverge significantly in content, the kinds of information they privilege, and the overall effect conveyed. The explanation is simple: like the differences between a small-scale and large-scale map, the scope and purpose of the Illustrations of Tributaries differed from that of a gazetteer, and of the albums. More than an aid in the governance of a specific region, it was to be a reflection that the various peoples portrayed had "turned toward civilization." The idea was to display the splendors of the realm in all their diversity. The imperial nature of the project also accounts for the attention to the administrative history of each group at the expense of some richness of detail. For the central government it was important to have a record of the provenance of each group, of what point it came under central authority, and in which precise jurisdiction. Used with care, the Illustrations of Tributaries constitute valuable historical source material for learning about how China's non-Han peoples and other peoples beyond the reaches of the Qing empire were categorized and understood by the those who had an interest in depicting them.

In their descriptions of peoples of Guizhou Province, the Illustrations of Tributaries contain some overlap with the albums. This can be seen in the Illustrations' concern with the history and origins of the various groups, their nature, customs, clothing, and the regions where they lived. Judging by the instructions issued in the commissioning edict quoted in chapter 1, these texts represent a shift during the Ming-Qing period toward reliance on direct observation and personal experience as the preferred methods of gathering practical information useful in the administration of frontier areas. They were based on observations made of foreigners who had visited China, or on information gathered when Chinese went abroad.[2] Yet, they exhibit less interest in accumulating a

2. The preface to the *Siku quanshu* edition also remarks that the Qing Imperial Tribute Illustrations produced under Qianlong differed from illustrations of tributaries of earlier dynasties in

great quantity of descriptive detail than many of the Miao albums. Instead they focus on recording the administrative history of each of the groups, and presenting an ordered and consistent picture of the peoples portrayed.

The Miao albums can be thought of as a much larger scale version of the Illustrations of Tributaries. Because they represented a correspondingly smaller geographic scope, more detailed representation was possible. They too represent various peoples of the empire, but in greater depth. The albums, at least their initial conception, served as informational documents aimed at a relatively limited audience: officials who desired as much information as possible about the groups described therein. Of most pressing concern were the present habits and customs of the different peoples who had recently come under their jurisdiction. Details, whether textual or pictorial, that would lead to identifying the different groups and signal what kind of behaviors and practices to expect of them was of primary importance. By contrast to the Illustrations of Tributaries, information on their history and provenance, while sometimes included, was less crucial. Illustrations that pictured multiple figures engaged in activities related to their beliefs, livelihood, or other distinctive practices served a purpose distinct from that of the Illustrations of Tributaries.

Thinking of these various representations of the Qing empire as different kinds of maps—maps of varying scale as well as maps representing both territory and peoples—points to an important characteristic of visual representation of empire in the early modern period. Peoples and territories, while both attracting closer scrutiny, increasingly came to be mapped in separate genres. Cartography and ethnography were in the process of diverging during this period to become separate disciplines.

In the sixteenth-century maritime atlases discussed in the introduction, indigenous peoples appear pictured in their own environment, often on the edge of the known world. Coastlines are marked "cartographically," but the interior of unfamiliar lands, which are pictured more pictorially, were at the juncture where the knowledge of the mapmaker (and viewer) began to fade into the unknown. Different peoples, customs, and habitats were worth learning about because they were unfamiliar; these lands and their peoples were a curiosity. In a sense it was precisely their difference that made their pictorial representation of

---

that the depictions were more reliable and true to life. Introduction (ti yao), 2. See also F. Jaeger, "Über chinesische Miaotse-Albums" (1917), 82–83.

interest and value on the map. In seventeenth-century European world maps, we continue to see an interest in the customs and costumes of other peoples, but the classification is more schematic or rigid. By the eighteenth century, maps were devoted exclusively to territorial features; descriptions of peoples instead appeared in other venues.

The early modern distinction between cartography and ethnography was more complete in Europe than under the Qing, where indigenous mapping practices were not superseded by the early modern practices put to use by the court. Nonetheless, in the limited but significant Qing use of scaled mapping described in this book, one also sees a disappearance of ethnographic features. At the same time the emergence of the Miao album as a separate ethnographic genre distinct from other broader geographical works shows evidence of the imprint of an outlook that distinguishes territory and the peoples who inhabit it.

Uncovering the use of ethnographic representation in Qing China as part of the process of empire building is important for its own sake, but it becomes even more significant when seen in the context of the use of broader visual representation internationally during the early modern period. When examined in this context the Qing use of both cartography and ethnography to define its territory and its peoples reveals an interconnectedness with the early modern world that has too long been overlooked. While maps and ethnographic representations can have a variety of different purposes, there can be no doubt that their use in defining Qing China as an emergent world force in the eighteenth century has parallels to the use of these types of visual representation in other parts of the world. In regard to ethnographic representation, whether we look at John White's representations of the New World, Japanese representations of the Ainu peoples, or Edward Lear's watercolors of India,[3] these are all attempts to gain information about other peoples. Early efforts were limited to obtaining knowledge about foreign peoples; later efforts often carried with them the goal of incorporating the peoples portrayed into the system of governance of those who did the depicting and classifying.

Mapping territory, whether by Portugal, France, Britain, Russia, Siam, or Qing China, was likewise carried out during this period in an effort to claim the physical extent, and later demarcate more strictly the boundaries, of emergent nation-states. No longer can Qing China,

---

3. Vidya Dehejia, *Impossible Picturesqueness: Edward Lear's Indian Watercolors, 1873–1875* (New York: Columbia University Press, 1989).

or the other countries of the early modern world, be seen in historical isolation from each other. They each participated in and responded to (sometimes differently) similar world forces.

As early modern maps helped to determine the territorial boundaries of the modern nation-state, ethnographies would help to define its citizens. An examination of the Miao albums and their prefaces uncovers a recurring theme of the need to better understand, to more thoroughly identify, and to better anticipate the natures and customs of the peoples in question—for administrative purposes. But more than this, along with the Qing Imperial Illustrations of Tributaries, the Miao albums were a step on the way to constructing ethnic diversity as a feature that helps to define the People's Republic of China today as seen in current picture books on China's minority nationalities.[4] For just as the album genre went through changes in content, presentation, and purpose over time, illustrated ethnographic representations continue to be produced about these same peoples today.

During a "minorities festival" held in September of 1991, large wooden placards illustrating China's southwestern minorities and detailing their dwelling places, population figures, and customs lined the main thoroughfare of the Yuan Ming Yuan, the ruins of a Qing summer palace located north of Peking. Further inside the park representatives of various non-Han groups manned exhibits and put on performances relating to their cultural heritage. This festival is an example of efforts on the part of the government of the People's Republic of China to create political unity in the face of ethnic and cultural diversity. By encouraging the participation of many different ethnic groups, the authorities used the festival as a forum for promoting the concept of national unity. At the same time, by creating a forum where Han Chinese could observe peoples from different parts of the country, the government made a statement about the extent and scope of its domain. The only major differences between the brightly colored placards and the albums to which they were so stylistically similar was the use of modern instead of classical Chinese, the broader audience they were designed to reach, and the inclusion of population figures in the details recorded. The placards

4. See, for example, *A Happy People—the Miaos* (Beijing: Foreign Languages Press, 1988). For more on present-day representations of "others" in the People's Republic of China, see Dru Gladney, "Representing Nationality in China: Refiguring Majority/Minority Identities," *Journal of Asian Studies* 53.1 (1994): 92–123; and Schein, "Gender and Internal Orientalism in China." On the construction of Muslim *(hui)* identity, see Dru Gladney, *Muslim Chinese: Ethnic Nationalism in the People's Republic* (Cambridge, Mass.: Council on East Asian Studies, Harvard University, 1991).

functioned not only to educate Chinese (and other) visitors to the summer palace about the customs and circumstances of the groups on display, they also served to claim these groups as part of the mix of peoples constituting the modern Chinese nation. In their content and composition the albums are not entirely different from some ethnographic materials that describe Guizhou's non-Han peoples today.

The reasons that Qing China did not continue to participate in patronizing cutting-edge world technology are diverse and complex. Although this question would form the basis for another book, it is fitting to suggest a few possibilities here. First, no single country has long remained a world leader in terms of either political dominance or technological innovation for very long. Even in Europe centers of power and influence shifted over time from Portugal to Italy to France to Britain. Rather than seeing Qing China's case as separate from these countries', it makes more sense to see it within the same chronology, peaking in its power and patronage of science at approximately the same time as Louis XIV's France. That France's monarchy fell at the end of the eighteenth century and Qing China's held on for another century could at an earlier time (for example, in 1793 when the Qianlong emperor spurned the Macartney mission sent by King George III) have been seen as a measure of Qing success in the face of a serious European disaster. If, as I will explore in another project, the monarchy measured itself against other monarchs internationally, France's decline may have discouraged Qing rulers from what they perceived as a route leading toward their own demise.

Another reason that the Qing may have chosen, or felt forced, to reject the opportunities that early modern technologies offered was that then, as now, they also carried the stigma of being "foreign." Even as early as the Kangxi emperor's reign, the throne felt pressure from Han Chinese to reject foreign influence. Jesuits, who provided a window on events in Europe and served as a conduit for early modern technology, also brought with them the potentially threatening doctrine of Christianity. Issues of "national sovereignty" came to be at stake. The rites controversy, in which the authority of the Pope over Chinese Catholic converts came into tension with the authority of the emperor, crystallized this threat and was responsible for the eventual disappearance of Jesuits from the court. The access to international channels for staying current and contributing to the latest developments—always crucial to leading centers of power—faded out along with their presence.

Other pressures, both internal and external, burdened the state. Internally, massive population growth and a series of ever more serious rebellions stretched the Qing to the breaking point. Foreign aggression in the form of the Opium War (1840–1842) and subsequent skirmishes intensified distrust of the West and what were seen as Western or foreign technologies.[5] Only in the 1860s when figures such as Wei Yuan began to recognize the necessity of using modern technology to defend China did Western learning begin to flourish more broadly.[6] However, in the nineteenth century the conditions were much different than they had been in the early eighteenth. Instead of being in a position to pick and choose what technologies might be useful, and to limit the extent of their circulation, China was forced increasingly (as it still is) to adopt modern technologies wholesale for survival. Then, as now, vocal groups continued to want to limit outside influence, or at least to use "Western" ways where useful but to keep the Chinese root at the core. Recent efforts to curtail "bourgeois liberalization" still reflect the same efforts to limit what is perceived as the pernicious influence of foreign ideologies.

Finally, we have to recognize that there is no preordained trajectory for the development of technologies and for the course that governments will pursue and histories take. What does become apparent through this study, however, is that from (at least) the early modern period, Qing China did not function in isolation from other parts of the world; the range of choices with which it was confronted and the situations it faced were shaped on the world stage. In the eighteenth century the Qing dynasty was in a position to choose to be involved in the forefront of empire building using early modern technologies of representation including ethnography and cartography as tools. That this was no longer possible in the nineteenth century is another story, but one that should not obscure or distort our understanding of earlier Qing engagement with the early modern world.

5. As James Polachek demonstrates, factionalism and Han Manchu tensions were also at issue in the Opium War. See James M. Polachek, *The Inner Opium War* (Cambridge, Mass.: Council on East Asian Studies, Harvard University, 1992).

6. See Jane Kate Leonard, *Wei Yuan and China's Rediscovery of the Maritime World* (Cambridge, Mass.: Council on East Asian Studies, Harvard University, 1984).

APPENDIX

The appendix lists bibliographic information for each of the Miao albums including their location and size. Albums are organized by province in the following order: Guizhou, Yunnan, Guangdong, Hunan, and Taiwan.

ALBUMS OF GUIZHOU PROVINCE

1. "Bashier zhang wanren tu" 八十二張萬人圖. Anonymous. 12.75 in. × 8.5 in. (actual illustration 8.25 in. × 7 in.). Musée Guimet, no. 3568. Eighty-two entries. Illustrations only, no text.

2. "Chou shi zhou xiansheng shanshui Miaojing renwu zhenji" 仇十州先生山水苗景人物真蹟. By Chou Ying. 13.5 in. × 9.5 in. American Museum of Natural History, New York, no. 44 (1) III-8. Twenty-four entries. Illustrations only, no text.

3. "Yongzheng yuzhi Miao tu ce shi kai" 雍正御製苗圖冊十開. Inscribed (ti 題) by Bing Liancheng 并連城. 10 in. × 11.25 in. American Museum of Natural History, New York, 381 (1) III-8. Ten entries. Commissioned during the Yongzheng reign period (1723–1735).

4. "Man liao tushuo" 蠻獠圖說. Anonymous magistrate of Guizhou, inscribed by Luo Lian 羅溓. 12 in. × 8.75 in. Bodleian Library, Oxford University, MS.Chin.c.15. Eighty-two entries, preface.

5. "Miao jiang tushuo" 苗疆圖說. Edited by Wang Jin 王鈞. 9.25 in. × 6 in. Bodleian Library, Oxford University, MS.Chin.d.20. Seventy-seven entries. Text only, no illustrations.

6. Untitled. Anonymous. 12.6 in. × 10.75 in. Museum für Völkerkunde, Abteilung Ostasien, Staatliche Museen Preussischer Kulturbesitz,

Berlin, no. 46973. Thirty-seven entries, preface, and postface. Preface dates album to 1768 (although this may be a later copy).

7. "Luo dian yi feng" 羅甸遺風, and "Nong sang ya hua" 農桑雅化, 2 vols. Anonymous. 11.5 in. × 8 in. British Library, Add. 16594 and Add. 16595. Vol. 1, twenty entries; vol. 2, twenty. Includes poetry.

8. Untitled. Anonymous. 9.5 in. × 5.5 in. British Library, Or. 11513. Twelve entries.

9. "Qian sheng zhu Miao shuo gong bashier zhong" 黔省諸苗説共八十二種. Anonymous. 10.5 in. × 6.5 in. British Library, Or. 11619. Eighty-two entries. Text only, no illustrations. Includes poetry.

10. "Guizhou quan qian Miao tu" 貴州全黔苗圖. Anonymous. 12.5 in. × 11 in. British Library, Or. 13504. Forty entries.

11. "Qian sheng ge zhong Miao tu" 黔省各種苗圖, 2 vols. Anonymous. 13.5 in. × 9.5 in. British Library, Or. 2232. Vol. 1, forty-one entries; vol. 2, thirty-seven .

12. "Miao tu" 苗圖. Anonymous. 13.75 in. × 11.75 in. British Library, Or. 4153. Forty-eight entries. Illustrations only, no text.

13. Untitled. Anonymous. 12 in. × 11.5 in. British Library, Or. 5005. Eighteen entries.

14. Untitled, 4 vols. Anonymous. 14 in. × 10.5 in. British Library, Or. 9623. Each volume contains eighteen entries.

15. "Bai Miao tu yong" 百苗圖永, 6 vols. Zou Yuanji 鄒元吉. Approx. 11 in. × 6.5 in. Guizhou Nationalities Research Institute, 012508-012513/K892.316/1988.11.22. Vol. 1, eighteen entries; vol. 2, eighteen; vol. 3, fourteen; vol. 4, sixteen; vol. 5, fourteen; vol. 6, twenty. Altogether one hundred entries, prefaces. Includes poetry. Dated 1890.

16. "Fan Miao hua ce" 番苗畫冊. Anonymous (includes falsified seal of Giuseppe Castiglione). 12 in. × 8.5 in. Institute of History and Philology, Academia Sinica. Sixteen entries. Includes poetry.

17. Untitled. Anonymous. 9.75 in. × 11 in. Museum für Völkerkunde, Leipzig (in Archives), FIII d1 (previously no. OAs 9864.China. Unbek). Thirty-six entries. Illustrations only, no text.

18. "Guizhou Miao tu" 貴州苗圖. Meng Shaojun 夢韶鈞. 14 in. × 9 in. Peking University Rare Book Room, NC2213/5346 50-867. Eighty-one entries, preface.

19. "Guizhou Miaozu tu" 貴州苗族圖. Anonymous. 15 in. × 11 in. Guizhou Provincial Museum. Sixty-five entries.

20. "Guizhou quan sheng Miaozu tushuo" 貴州全省苗族圖説. Anonymous. 11.5 in. × 8.25 in. National Library, Peking, 041.3 (236) 1820 9017. Forty-one entries.

21. "Guizhou quan sheng zhu Miao tushuo" 貴州全省諸苗圖説. Anonymous. 9.5 in. × 8.5 in. Beihai branch of National Library, Peking, Ce 041.3 (236) 1820 9017 (this will change; the book was being reclassified in 1991). Eighty-two entries. Text only, no illustrations.

22. "Guizhou xiongdi minzu tu" 貴州兄弟民族圖. Seals may indicate authorship. 10 in. × 12 in. (actual illustrations 8 in. × 10 in.). National Library, Peking, Ce 041.30 (236) 1900 10481. Seventeen entries. Illustrations only, no text. Individual drawings, not bound.

23. "Guizhou xiongdi minzu tu" 貴州兄弟民族圖. Seals may indicate authorship. 10.5 in. × 8.25 in. National Library, Peking, Ce 041.3 (236) 1850 10468. Forty entries.

24. "Guizhou xiongdi minzu tu" 貴州兄弟民族圖. Seals may indicate authorship. 10 in. × 7.5 in. National Library, Peking, Ce 041.3 (236) 1850 10468. Forty entries.

25. Untitled. Anonymous. 11.25 in. × 8 in. Harvard-Yenching Treasure Room, uncataloged. Forty-one entries. Includes poetry.

26. "Qian sheng Miao tu" 黔省苗圖. Seals may indicate authorship. 9.85 in. × 10.5 in. Società Geografica Italiana, Chinese Catalog no. 57. Eighteen entries.

27. Untitled (title worn off, but the character *shang* (上) indicates this may be the first volume of a multivolume set). Anonymous. 12.5 in. × 9.1 in. Società Geografica Italiana, Chinese Catalog no. 59. Seventy-two entries. Illustrations only, no text.

28. "Bai Miao tu" 百苗圖. Anonymous. 10.75 in. × 12.15 in. Società Geografica Italiana, Chinese Catalog no. 60. One hundred entries. Includes poetry.

29. Untitled (original title worn off). Seals may indicate authorship. 11 in. × 12.5 in. Società Geografica Italiana, Chinese Catalog no. 61. Forty entries.

30. "Qiansheng Miao tu quanbu" 黔省苗圖全部. Anonymous. 12.75 in. × 8.15 in. Società Geografica Italiana, Chinese Catalog no. 63. Eighty entries. Includes poetry.

31. Untitled. Anonymous. 10 in. × 7.5 in. Società Geografica Italiana, Chinese Catalog no. 66. Sixty-seven entries.

32. Untitled. Jiao Bingzhen 焦秉真. 14.25 in. × 14 in. (actual illustrations 9.75 in. × 12.5 in.). Società Geografica Italiana, Chinese Catalog no. 68. Twelve entries. Illustrations only, no text.

33. Untitled. Seals may indicate authorship. 13.5 in. × 9.4 in. Società Geografica Italiana, Chinese Catalog no. 71. Sixteen entries.

34. Untitled. Anonymous. Scroll: 90 in. × 27 in. (actual illustration 61 in.

× 19.5 in.). Società Geografica Italiana, Chinese Catalog no. 72. Ten entries.

35. "Miaofeng tulu" 苗風圖錄. Anonymous. 10 in. × 6.5 in. Peking University Library, Rare Book Room. Seventy-two entries.

36. "Miaoman fengsu tuzhi" 苗蠻風俗圖誌. Anonymous. 7.5 in. × 5 in. Peking University Library, Rare Book Room, no. 54211. Seventy-eight entries.

37. "Miao man tu" 苗蠻圖. Anonymous. Central Nationalities Institute Library, no. 153831. Fifty-nine entries.

38. "Miao man tu ce" 苗蠻圖冊. Anonymous. 11 in. × 12 in. Institute of History and Philology, Academia Sinica, no. 132276. Eighty-two entries.

39. "Miao man tu" 苗蠻圖. Zhou Zhimian 周之冕. 12.5 in. × 9 in. Institute of History and Philology, Academia Sinica. Twenty-seven entries, preface.

40. "Miao man tu" 苗蠻圖. Seals may indicate authorship. 10 in. × 10 in. Institute of History and Philology, Academia Sinica, no. 129318. Twenty-five entries.

41. "Miao man tu ce ye" 苗蠻圖冊頁. Fang Ting 舫亭. 12.5 in. × 9 in. Library of Congress, East Asian Collection, Rare Book, D827 M59. Forty-one entries.

42. "Miao man tushuo" 苗蠻圖説. Anonymous. 13 in. × 9.75 in. Harvard-Yenching Library, Treasure Room. Forty-two entries.

43. "Miao man tushuo" 苗蠻圖説 (title on outer cover box only). Mu Konggong 木孔恭. 10.5 in. × 11 in. Harvard-Yenching Library, Treasure Room. Thirty-seven entries.

44. "Miao man tushuo" 苗蠻圖説, ba er kai 八二開. Anonymous. 9.75 in. × 11.5 in. Harvard-Yenching Library, Treasure Room. Eighty-two entries.

45. Untitled. Anonymous. 10 in. × 6 in. Museum of Natural History, Washington, D.C., accession number 258148. Forty-two entries.

46. "Mingren jingxie Miao man tu" 名人精寫苗蠻圖, 2 vols. Anonymous. 12 in. × 7.5 in. Forschungs- und Landesbibliothek Gotha, Ms. orient. Ag 17a and 17b. Vol. 1, forty-one entries; vol. 2, forty.

47. "Miao tu liushisi ye" 苗圖六十四頁. Anonymous. 12.5 in. × 8.75 in. Musée Guimet, no. 23244. Sixty-four entries.

48. "Miao Yao zu shenghuo tu" 苗猺族生活圖. Anonymous. 14 in. × 11.5 in. Princeton University, Gest Library, Rare Book Room, C-223 Gest No. 2146. Forty entries.

49. "Miaozu fengsu tushuo" 苗族風俗圖説, 2 vols. Anonymous. Central

Nationalities Institute, Library, *Xian* 459.203 0.523 (1-2). Seventy-nine entries (incomplete).

50. "Qiansheng Miao tu" 黔省苗圖, 4 vols. Museum catalog names Wang Jin 王鈞 as author, but the album itself gives no such indication. 9.5 in. × 5.75 in. Pitt Rivers Museum, Oxford, no. 1917.53.723.1-4. Vol. 1, twenty entries; vol. 2, twenty-two; vol. 3, twenty; vol. 4, twenty.

51. Untitled, 3 vols. Museum catalog names Wang Jin 王鈞 as author, but the album itself gives no such indication. 14 in. × 11 in. Pitt Rivers Museum, Oxford, no. 1917.53.724.1-3. Vol. 1, twelve entries (only eleven illustrations); vol. 2, nineteen (missing one page of text); vol. 3, eighteen.

52. "Qian Miao tu" 黔苗圖. Anonymous. 10.75 in. × 8.5 in. Institute of History and Philology, Academia Sinica, no. 85360. Eighty-two entries. Illustrations only, no text.

53. "Qian Miao tu" 黔苗圖. Anonymous. 12 in. × 12 in. (actual illustration 8 in. × 10.5 in.). Institute of History and Philology, Academia Sinica, no. 85362. Eighty-two entries. Illustrations only, no text.

54. "Qian Miao Tushuo" 黔苗圖説. Anonymous. 12.5 in. × 8.75 in. Chinese Academy of Social Sciences, Nationalities Research Institute, Peking. Eighty two entries

55. "Qian Miao tushuo" 黔苗圖説. Guo Peiyuan 郭培元. Guizhou Provincial Museum. Eighty entries.

56. "Qian Miao tushuo" 黔苗圖説. Anonymous. 13 in. × 8.75 in. Harvard-Yenching Library, Treasure Room. Fifty-seven entries. Includes poetry.

57. "Qian Miao tushuo" 黔苗圖説. Anonymous. Institute of History and Philology, Academia Sinica. Eighty entries.

58. "Qian Miao tushuo bu" 黔苗圖説補. Anonymous. 11.5 in. × 9.5 in. (actual illustrations 6.5 in. × 7.5 in.). Institute of History and Philology, Academia Sinica, no. 85355. Seven entries. Illustrations only, no text.

59. "Qiannan bashier zhong Miao tu" 黔南八十二種苗圖. Anonymous. Guiyang Wenwu Shangdian (not for sale). Eighty-two entries.

60. "Qiannan Miao man tushuo" 黔南苗蠻圖説. Some texts are followed by a signature. 11.5 in. × 10 in. Central Nationalities Institute, Nationalities Reading Room. Eighty-six entries, prefaces. Illustrations are black ink on white rice paper.

61. "Quan qian Miao tu" 全黔苗圖. Anonymous. 12 in. × 10.5 in. The University Museum, University of Pennsylvania. Forty entries.

62. "Quansheng Miaozu shenghuo" 全省苗族生活. Anonymous. 12.5 in.

× 8.5 in. Chinese Academy of Social Sciences, Nationalities Research Institute, Peking. Eighty-two entries. Includes poetry.

63. "Qiansheng zhu Miao shuo" 黔省諸苗説. Anonymous. 13 in. × 8 in. New York Public Library, Spencer Collection, manuscript Chinese 10. Eighty-one entries.

64. Untitled. Anonymous. Scroll: 53 in. × 13.6 in. Museum für Völkerkunde, Leipzig, OAs 13298 China.Konietzko. Scroll contains two illustrations with accompanying text, and blank space for a third.

65. "Zhongguo neidi fan Miao fengsu tu" 中國內地番苗風俗圖. Anonymous. 14.25 in. × 15.25 in. Musée Guimet, no. 32216. Twenty-six entries.

Albums of Yunnan Province

66. "Pu'er fu yudi yiren tushuo" 普洱府輿地夷人圖説. Anonymous. 14.75 in. × 12 in. British Library, Or. 6588. Thirteen entries, map, preface.

67. "Diansheng yiren tushuo" (shang) 滇省夷人圖説(上), and "Diansheng yudi tushuo" (xia) 滇省輿地圖説(下). Anonymous. Chinese Academy of Social Sciences, Nationalities Research Institute, Peking.

68. Untitled. Anonymous. 12.5 in. × 9.5 in. Società Geografica Italiana, Chinese Catalog no. 64. Fifty-four entries.

69. "Dian yi tushuo" 滇夷圖説, 4 vols. Anonymous. 15 in. × 15 in. Institute of History and Philology, Academia Sinica. Each volume contains twelve entries (vol. 1 is missing one page of text) and one map.

70. "Miao man tu" 苗蠻圖. Anonymous. 20 in. × 12 in. Chinese Academy of Social Sciences, Nationalities Research Institute, Peking.

71. "Yongbei diyu bing tusi suoshu yiren zhonglei tu"* 永北地輿並土司所屬夷人種類圖. Anonymous. 10.5 in. × 8.75 in. Società Geografica Italiana, Chinese Catalog no. 70. Six entries, map.

72. "Yunnan san yi bai man tu quanbu" 云南三地百蠻圖全部, 4 vols. 14.75 in. × 11.5 in. Bodleian Library, Oxford University, Ms.Chin.c.37. Vols. 1 and 3 include maps and explanatory text, vols. 2 and 4 include ethnographic illustrations with accompanying text. Postface.

73. "Yunnan yi lei tu" 云南夷類圖 (on outside cover), "Yunnan liang yi yi lei tushuo" 云南兩地夷類圖説 (on inside). Li Ji 禮齊. 13 in. × 10 in. British Library, Or. 4152. Forty-four entries, introduction, table of contents.

* An archival sticker covers bottom of last character (tu) in title. Impossible to tell if another entire character (shuo 説) is obscured or not.

74. "Dian Miao tushuo" 滇苗圖説, 2 vols. Anonymous. 16 in. × 11 in. Harvard-Yenching Library, Treasure Room. Each volume contains eighteen entries.

75. "Dian sheng yi xi yi nan yiren tushuo" 滇省迤西迤南夷人圖説. He Changgeng 賀長庚. 12.5 in. × 11 in. Museum für Völkerkunde, Leipzig (archives). F III d 2. Forty-four entries, preface.

76. Untitled. Anonymous. 16.5 in. × 10.25 in. Società Geografica Italiana, Chinese Catalog no. 58. Fifty-three entries.

77. Untitled. Anonymous. 12.5 in. × 9.5 in. Società Geografica Italiana, Chinese Catalog no. 65. Fifty-four entries.

78. "Miao fan tu" 苗蕃圖, 2 vols. Seals may indicate authorship. 12.5 in. × 8.75 in. Library of Congress, Orientalia Chinese D827 M58. Each volume contains eighteen entries, but vol. 2 contains illustrations only, no text.

ALBUM OF GUANGDONG PROVINCE

79. "Lianshanting Lianzhou fen xia Yao pai diyu quantu" 連山廳連州分轄猺排地輿全圖. Anonymous. 11.5 in. × 6.5 in. Società Geografica Italiana, Chinese Catalog nos. 62 and 67. Nine entries, large foldout map.

HAINAN ISLAND, GUANGDONG

80. "Liren fengsu tu" 黎人風俗圖. Anonymous. 17 in. × 13 in. Institute of History and Philology, Academia Sinica, no. 129605. Eighteen entries, preface.

ALBUM OF HUNAN PROVINCE

81. "Chunan Miaojiang tushuo" 楚南苗疆圖説. Anonymous. 12.75 in. × 8.75 in. Musée Guimet, no. 32229. Thirteen entries, but only twelve illustrations.

ALBUM OF TAIWAN

82. "Taifan tushuo" 台番圖説. Anonymous. 14 in. × 11 in. Institute of History and Philology, Academia Sinica. Eighteen entries, map.

ABBREVIATIONS

GZTZ	*Guizhou tongzhi* (1673, 1692, or 1741 edition as indicated).
QSLSZ	*Daqing lichao shilu: Qing Shengzu chun huangdi shilu.*
QSLGZ	*Daqing lichao shilu: Qing Gaozong chun huangdi shilu*
SKQS	*Siku quanshu*

BIBLIOGRAPHY OF WORKS CITED

For Miao albums, see Appendix

Adas, Michael. "Imperialism and Colonialism in Comparative Perspective." *International History Review* 20.2 (June 1988): 371–88.

Allgemeine deutsche Biographie. Prepared by the Historical Commission of the Royal Academy of Sciences, Bavaria. 1886. Reprint, Berlin: Duncker & Humblot, 1970.

Les animaux et les arts. Includes Claude Perrault's *Memoires pour servir à l'histoire naturelle des animaux.* Paris: Imprimerie Royale, 1671–1676, and plates made for use in the *Descriptions des arts et métiers* of the Académie des Sciences, Paris, 1761–1789.

Anderson, Benedict. *Imagined Communities: Reflections on the Origin and Spread of Nationalism.* Revised edition. London and New York: Verso, 1991.

Asad, Talal. "The European Images of non-European Rule." In *Anthropology and the Colonial Encounter.* London: Ithaca Press, 1973.

———, ed. *Anthropology and the Colonial Encounter.* London: Ithaca Press, 1973.

Backus, Charles. *The Nan-Chao Kingdom and T'ang China's Southwestern Frontier.* Cambridge and New York: Cambridge University Press, 1981.

Baddeley, John F. "Father Matteo Ricci's Chinese World Maps, 1584–1608." *Geographical Journal* 50 (1917): 254–70.

———. *Russia, Mongolia, and China.* New York: Burt Franklin, 1919.

Barr, Pat. *To China with Love: The Lives and Times of Protestant Missionaries in China, 1860–1900.* New York: Doubleday and Co., 1973.

Beauclair, Inez de. *Tribal Cultures of Southwest China.* Asian Folklore and Social Life Monographs, vol. 2. Edited by Lou Tsu-k'uang in collaboration with Wolfram Eberhard. Taipei: Orient Cultural Service, 1970. (Consists of reprints of earlier articles.)

Berliner Festspiele. *Europa und die Kaiser von China.* Frankfurt am Main: Insel Verlag, 1985.

———. *Japan und Europa, 1543–1929.* Berlin: Argon Verlag, 1993.

Bertuccioli, Giuliano. "Chinese Books from the Library of the Italian Geographical Society in Rome Illustrating the Lives of Ethnic Minorities in Southwest China." *East and West* 37 (1987): 399–438.

Beurdeley, Cecile and Michel. *Giuseppe Castiglione: A Jesuit Painter at the Court of the Chinese Emperors.* Rutland, Vt., and Tokyo: Charles E. Tuttle Company, 1971.

Biographical Dictionary of America, The. Boston: American Biographical Society, 1906.

Birrell, Anne. *Chinese Mythology: An Introduction.* Baltimore and London: Johns Hopkins University Press, 1993.

Bräutigam, Herbert. "Über Miao-Alben." *Mitteilungen des Instituts für Orientforschung* 9 (1963): 284–309.

Bremen, Jan van, and Akitoshi Shimizu, eds., *Anthropology and Colonialism in Asia and Oceania.* Richmond, Surrey: Curzon Press, 1999.

Bridgman, E. C. "Sketches of the Miao-tsze." *Journal of the North China Branch of the Royal Asiatic Society* 1.2 (1859): 257–86.

Brokaw, Cynthia J. "Commercial Publishing in Late Imperial China: The Zou and Ma Family Businesses." *Late Imperial China* 17.1 (1996): 49–92.

Brown, Lloyd, A. *The Story of Maps.* New York: Dover Publications, 1979.

Brunnert, H. S., and V. V. Hagelstrom. *Present Day Political Organization of China,* translated by A. Beltchenko and E. E. Moran. Shanghai: Kelly and Walsh, 1912.

Buisseret, David. "Monarchs, Ministers, and Maps in France before the Accession of Louis XIV." In *Monarchs, Ministers, and Maps: The Emergence of Cartography as a Tool of Government in Early Modern Europe.* Chicago and London: University of Chicago Press, 1992.

———, ed. *Monarchs, Ministers, and Maps: The Emergence of Cartography as a Tool of Government in Early Modern Europe.* Chicago and London: University of Chicago Press, 1992.

Bushell, Stephen W. *Chinese Art.* 2 vols. London: Printed under the Authority of His Majesty's Stationery office, 1909.

Cao Wanru, et al., eds. *Zhongguo gudai ditu ji* [A Collection of Ancient Chinese Maps]. Vol. 3, *Qing dai* [the Qing dynasty]. Peking: Wenwu chubanshe, 1997.

Cartier, Michel. "Barbarians through Chinese Eyes: The Emergence of an Anthropological Approach to Ethnic Differences." *Comparative Civilization Review* 6 (1981): 1–14.

Ch'en, Kenneth. "Matteo Ricci's Contribution to, and Influence on, Geographical Knowledge in China." *Journal of the American Oriental Society* 59 (1939): 325–59.

———. "A Possible Source for Ricci's Notices on Regions Near China." *T'oung Pao* 34 (1938): 179–90.

Chia Ning. "The Manchu Rule in Early Modern Mongolia: Religious or Secular?" Paper presented at the annual meeting of the Association for Asian Studies, March 26–29, 1998.

———. "Mongol Tribute in Seventeenth- and Eighteenth-Century China: The Significance of Gift Exchange in the Formation of National Boundaries." Paper presented at the annual meeting of the American Historical Association, January 5–8, 1995.

———. "The Lifanyuan and the Inner Asian Rituals in Early Qing (1644–1795)." *Late Imperial China* 14.1 (1993): 60–92.

Chiu Chang-kong. "Die Kultur der Miao-tse nach älteren chinesischen Quellen." *Mitteilungen aus dem Museum für Völkerkunde in Hamburg* 18 (1937): 5–32.

Chu, Pingyi. "Trust, Instruments, and Cross-Cultural Scientific Exchanges: Chinese Debate Over the Shape of the Earth, 1600–1800." *Science in Context* 12.3 (1999): 385–411.

Chuang Chi-fa. *Xie Sui "zhigong tu" manwen tushuo jiaozhu.* Taipei: Guoli gugong bowuyuan, 1989.

———. "Xie Sui zhigong tu yanjiu" [A Study of the *Tribute Presenting Scroll* by Hsieh Sui]. Proceedings of the 1991 Taipei Art History Conference, National Palace Museum, Taipei, 1992.

Churkin, V. G. "Atlas Cartography in Prerevolutionary Russia." In Bernard V. Gutsell, ed., *Essays on the History of Russian Cartography, 16th to 19th Centuries,* translated by James R. Gibson. Toronto: University of Toronto Press, 1975.

Cipolla, Carlo M. *Clocks and Culture, 1300–1700.* New York and London: W. W. Norton and Company, 1978.

Clarke, George W. "Translation of a Manuscript Account of the Kwei-chau Miao-tzu." In *Across Chryse,* by Archibald R. Colquhoun. London: Sampson Low, Marston Searle and Rivington, 1883.

Clarke, Samuel R. *Among the Tribes of Southwest China.* London: Morgan and Scott, 1911. Reprint, Taipei: Ch'eng-wen, 1970.

Clifford, James. *The Predicament of Culture: Twentieth-Century Ethnography, Literature, and Art.* Cambridge and London: Harvard University Press, 1988.

Clifford, James, and George E. Marcus, eds. *Writing Culture.* Berkeley, Los Angeles, and London: University of California Press, 1986.

Cohen, Paul A. *Discovering History in China.* New York: Columbia University Press, 1984.

Cohn, Bernard S. "The Census, Social Structure and Objectification in South Asia." In *An Anthropologist among Historians and Other Essays.* Delhi: Oxford University Press, 1987.

Colquhoun, Archibald. *Across Chryse.* 2 vols. London: Sampson Low, Marston Searle and Rivington, 1883.

Cracraft, James. *The Petrine Revolution in Imagery.* Chicago: University of Chicago Press, 1997.

Crosby, Alfred W. *The Measure of Reality: Quantification and Western Society, 1250–1600.* Cambridge: Cambridge University Press, 1997.

Crossley, Pamela Kyle. "*Manzhou yuanliu kao* and the Formalization of the Manchu Heritage." *Journal of Asian Studies* 46.4 (1987): 761–90.

———. *Orphan Warriors: Three Manchu Generations and the End of the Qing World.* Princeton: Princeton University Press, 1990.

———. *A Translucent Mirror: History and Identity in Qing Imperial Ideology.* Berkeley and Los Angeles: University of California Press, 1999.

Crossley, Pamela Kyle, and Evelyn S. Rawski. "A Profile of the Manchu Language in Ch'ing History." *Harvard Journal of Asiatic Studies* 53.1 (June 1993): 63–102.

Cushman, Richard David. "Rebel Haunts and Lotus Huts: Problems in the Ethnohistory of the Yao." Ph.D. diss., Cornell University, 1970.

Da Qing Gaozong Chun huangdi shilu [Veritable Records of the Qing Dynasty for the Qianlong Emperor's Reign]. Taipei: Hualian chubanshe edition, 1964.

Da Qing Shengzu Ren huangdi shilu [Veritable Records of the Qing Dynasty for the Kangxi Emperor's Reign]. Taipei: Hualian chubanshe, 1964.

Dai Yi, ed. *Jian ming Qing shi* [A Concise History of the Qing]. Renmin chubanshe, 1980.

Dainville, François de. *La géographie des humanistes.* Paris: Beauchesne et Ses Fils, 1940.

Dehejia, Vidya. *Impossible Picturesqueness: Edward Lear's Indian Watercolors, 1873–1875*. New York: Columbia University Press, 1989.

D'Elia, Pasquale M. "Il mappamondo Ricciano nell'atlante de Cuozzimluo (Kweiyang, Kweichow, 1604)." In *Il mappamondo cinese del P. Matteo Ricci S.I.*, 73–81. Città del Vaticano: Biblioteca Apostolica Vaticana, 1938.

———. *Il mappamondo cinese del P. Matteo Ricci S.I.* Città del Vaticano: Biblioteca Apostolica Vaticana, 1938.

———. "Recent Discoveries and New Studies on the World Map in Chinese of Father Matteo Ricci, S.J." *Monumenta Serica* 20 (1961): 82–164.

Dessaint, Alain Y. *Minorities of Southwest China: An Introduction to the Yi (Lolo) and Related Peoples and an Annotated Bibliography*. New Haven: HRAF Press, 1980.

Destombes, Marcel. "Une carte chinoise du XVIe siècle découverte à la Bibliothèque Nationale." In *Marcel Destombes: Contributions sélectionnées à l'histoire de la cartographie et des instruments scientifiques*, edited by Günter Schilder, Peter van der Krogt, and Steven de Clercq, 461–80. Utrecht: HES Publishers; Paris: A.G. Nizet, 1987.

———. "Wang P'an, Liang Chou et Matteo Ricci: Essai sur la cartographie chinoise de 1593 à 1603." In *Appréciation par l'Europe de la tradition chinoise: À partir du XVIIe siècle: Actes du IIIe Colloque International de Sinologie*, 47–65. Paris: Les Belles Lettres, 1983.

Di Cosmo, Nicola. "Qing Colonial Administration in Inner Asia." *International History Review* 20.2 (June 1988): 287–309.

Diamond, Norma. "The Miao and Poison: Interactions on China's Southwest Frontier." *Ethnology* 27 (1988): 1–25.

———. "Defining the Miao." In *Cultural Encounters on China's Ethnic Frontier*, edited by Stevan Harrell. Seattle and London: University of Washington Press, 1995.

Dictionary of American Biography. Vol. 3. New York: Charles Scribner's Sons, 1929.

Dikötter, Frank. *The Discourse of Race in Modern China*. London: C. Hurst, 1992.

Doggett, Rachel, ed., with Monique Hulvey and Julie Ainsworth. *New World of Wonders: European Images of the Americas, 1492–1700*. Washington, D.C.: Folger Shakespeare Library; Seattle: distributed by University of Washington Press, 1992.

Du Halde, Jean Baptiste. *Description géographique, historique, chronologique, politique, et physique de l'empire de la Chine et de la Tartarie chinoise, enrichie des cartes générales et particulières de ces pays, de la carte générale & des cartes particulières du Thibet, & de la Corée, & ornée d'un grand nombre de figures et de vignettes gravées en taille-douce*. 4 vols. Paris: P. G. Lemercier, 1735.

Duara, Prasenjit. *Rescuing History from the Nation: Questioning Narratives of Modern China*. Chicago and London: University of Chicago Press, 1995.

Dunn, Oliver, and James E. Kelley, Jr., trans. *The "Diario" of Christopher Columbus's First Voyage to America, 1492–1493*. Norman, Okla., and London: University of Oklahoma Press, 1989.

Dunne, George H., S.J. *Generation of Giants: The Story of the Jesuits in China in the Last Decades of the Ming Dynasty*. Notre Dame, Ind.: University of Notre Dame Press, 1962.

Duyvendak, J. J. L. Review of *The Mongol Atlas of China by Chu Ssü-pen and the Kuang-yü-t'u*, by Walter Fuchs. *T'oung Pao* 39 (1949): 197–99.

Eberhard, Wolfram. *China's Minorities: Yesterday and Today*. Belmont, Calif.: Wadsworth, 1982.

———. "Die Miaotse Alben des Leipziger Völkermuseums." *Archiv für Anthropologie.* 26 (1941): 125–37.

Edkins, J. "The Miau Tsi Tribes: Their History." *The Chinese Recorder and Missionary Journal* 2 (July 1870): 33–36.

———. "The Miau Tsi." *The Chinese Recorder and Missionary Journal* 2 (August 1870): 74–76.

Elman, Benjamin. "Geographical Research in the Ming-Ch'ing Period." *Monumenta Serica* 35 (1981–1983): 1–18.

Elman, Benjamin, and Alexander Woodside, eds. *Education and Society in Late Imperial China, 1600–1900.* Berkeley: University of California Press, 1994.

Fairbank, John King. *Trade and Diplomacy on the China Coast: The Opening of Treaty Ports, 1842–1854.* Cambridge, Mass.: Harvard University Press, 1948.

Farquhar, David M. "Emperor as Bodhisattva in the Governance of the Ch'ing Empire." *Harvard Journal of Asiatic Studies* 38 (1978): 5–34.

Fel, S. Ye. "The Role of Petrine Surveyors in the Development of Russian Cartography During the 18th Century." In Bernard V. Gutsell, ed., *Essays on the History of Russian Cartography, 16th to 19th Centuries,* translated by James R. Gibson. Toronto: University of Toronto Press, 1975.

Fenfu zhongguo quantu [Maps of the Provinces of China]. 1721. British Library, OIOC 15271.a.20.

Feng, H. Y., and J. K. Shryock. "The Black Magic Known in China as Ku." *Journal of the American Oriental Society* 55 (1935): 1–30.

Foss, Theodore N. "A Jesuit Encyclopedia for China: A Guide to Jean-Baptiste Du Halde's *Description . . . de la Chine* (1735)." Ph.D. diss., University of Chicago, 1979

———. "A Western Interpretation of China: Jesuit Cartography." In *East Meets West: The Jesuits in China, 1582–1773,* edited by Charles E. Ronan, S.J., and Bonnie B. C. Oh, 209–51. Chicago: Loyola University Press, 1988.

Foucault, Michel. *The Order of Things: An Archaeology of the Human Sciences.* New York: Vintage Books, 1973.

Frank, Andre Gunder. *Reorient: Global Economy in the Asian Age.* Berkeley: University of California Press, 1998.

Friedman, Jonathan. "Narcissism, Roots and Postmodernism: The Construction of Selfhood in the Global Crisis." In *Modernity and Identity,* edited by Scott Lash and Jonathan Friedman, 331–66. Oxford, U.K., and Cambridge, Mass.: Basil Blackwell, 1992.

Fu, Lo-shu. *A Documentary Chronicle of Sino-Western Relations, 1644–1820.* Association for Asian Studies: Monographs and Papers, no. 22. Tucson: University of Arizona Press, 1966.

Fuchs, Walter. *Der Jesuiten-Atlas der Kanghsi-Zeit: China und die Aussenländer.* Monumenta Serica, monograph no. 3. Peiping: Fu Jen University Press, 1941.

———. *Der Jesuiten-Atlas der Kanghsi-Zeit: Seine Entstehungsgeschichte nebst Namensindices für die Karten der Mandjurei, Mongolei, Ostturkestan und Tibet, mit Wiedergabe der Jesuiten-Karten in Original Grösse.* Monumenta Serica, monograph no. 4. Peiping: Fu Jen University Press, 1943.

———. *The "Mongol Atlas" of China, by Chu Ssu-Pen, and the "Kuang Yü Thu,"* Monumenta Serica, monograph no. 8. Peiping: Fu Jen University Press, 1946.

Ganza, Kenneth. "To Hear with the Ears Is Not as Good as to See with the Eyes: Travel

Painting as an Expression of Empiricism in the Late Ming and Early Qing." Conference paper presented at the annual meeting of the Association for Asian Studies, March 12–16, 1997.

Gaubil, Antoine, S.J. *Correspondance de Pékin, 1722–1759.* Geneva: Librairie Droz, 1970.

Giersch, C. Patterson. "Mapping an Imperial Frontier into 'National Territory': How Qing Officials Demarcated the Yunnan-Burma Frontier and Helped Produce a Corner of China." Paper presented at the annual meeting of the Association for Asian Studies, March 26–29, 1998.

Giles, Lionel. "Translations from the Chinese World Map of Father Ricci." *Geographical Journal* 52.6 (1918): 367–85.; 53.1 (1919): 19–30.

Gladney, Dru. *Muslim Chinese: Ethnic Nationalism in the People's Republic.* Cambridge, Mass.: Council on East Asian Studies, Harvard University, 1991.

———. "Representing Nationality in China: Refiguring Majority/Minority Identities." *Journal of Asian Studies* 53.1 (1994): 92–123.

Goldstone, Jack A. "The Problem of the Early Modern World." *Journal of the Economic and Social History of the Orient* 41 (August 1998): 249–84.

———. "Whose Measure of Reality?" Review of *The Measure of Reality: Quantification and Western Society, 1250–1600,* by Alfred W. Crosby. *The American Historical Review* 105.2 (April 2000): 501–8.

Goodall, John A. *Heaven and Earth: 120 Album Leaves from a Ming Encyclopedia, San-ts'ai t'u-hui, 1610.* London: Lund Humphries, 1979.

Goodman, Howard L. "Paper Obelisks." In *Rome Reborn: The Vatican Library and Renaissance Culture,* edited by Anthony Grafton, 251–91. Washington, D.C.: Library of Congress, in association with the Biblioteca Apostolica Vaticana, 1993.

Goodrich, L. Carrington, ed., and Chaoying Fang, assoc. ed. *Dictionary of Ming Biography, 1368–1644.* The Ming Biographical History Project of the Association for Asian Studies. New York and London: Columbia University Press, 1976.

La grande encyclopédie: Inventaire raisonné des sciences, des lettres et des arts. 31 vols. Paris: Société Anonyme de la Grande Encyclopédie, 1902.

Grasset-Saint-Sauveur, Jacques. *Tableaux des principales peuples de l'Europe, de l'Asie, de l'Afrique, de l'Amérique, et les découvertes des capitaines Cook, La Pérouse, etc. etc.* Paris, 1798.

Guizhou sheng dili. Edited by Guizhou Normal University, Department of Geography. Guiyang, Guizhou: Guizhou renmin chubanshe, 1990.

Guizhou Tongzhi. 1673 edition. National Library, Peking.

Guizhou Tongzhi. 1692 edition. Bibliothèque Nationale, Paris.

Guizhou Tongzhi. 1741 edition. Taipei: Jinghua shuju, 1968. Facsimile reproduction of 1741 (Qianlong 6 year) edition.

Guo Zizhang. *Qian ji.* 1608. Guizhou Provincial Library.

Gutsell, Bernard V., ed. *Essays on the History of Russian Cartography, 16th to 19th Centuries.* Translated by James R. Gibson. Toronto: University of Toronto Press, 1975.

Hansen, Valerie. *The Open Empire: A History of China to 1600.* New York and London: W. W. Norton and Company, 2000.

Happy People—the Miaos, A. Beijing: Foreign Languages Press, 1988.

Hariot, Thomas. *A Briefe and True Report of the New Found Land of Virginia. . . .* Manchester: The Holbein Society, 1888. Reproduction of the 1590 edition printed at Frankfort.

Harley, J. B., and David Woodward, eds. *The History of Cartography.* Vol. 2, bk. 2, *Cartography in the Traditional East and Southeast Asian Societies.* Chicago and London: University of Chicago Press, 1994.

Harrell, Stevan. "The History of the History of the Yi." In *Cultural Encounters on China's Ethnic Frontier,* edited by Stevan Harrell, 63–91. Seattle and London: University of Washington Press, 1995.

Hart, Roger. "The Great Explanadum." Review of *The Measure of Reality: Quantification and Western Society, 1250–1600,* by Alfred W. Crosby. *The American Historical Review* 105.2 (April 2000): 486–93.

Harvey, P. D. A. *The History of Topographic Maps: Symbols, Pictures, and Surveys.* London: Thames and Hudson, 1980.

Hechter, Michael. *Internal Colonialism: The Celtic Fringe in British National Development, 1536–1966.* Berkeley and Los Angeles: University of California Press, 1975.

Henderson, John B. "Chinese Cosmographical Thought: The High Intellectual Tradition." In *The History of Cartography,* vol. 2, bk. 2, *Cartography in the Traditional East and Southeast Asian Societies,* edited by J. B. Harley and David Woodward, 203–27. Chicago and London: University of Chicago Press, 1994.

———. *The Development and Decline of Chinese Cosmology.* New York: Columbia University Press, 1984.

Herman, John. "The Cant of Conquest: Creating 'Barbarians' and 'Chinese' in the Southwest." Conference paper presented at the annual meeting of the Association for Asian Studies, March 12–16, 1997.

———. "Empire in the Southwest: Early Qing Reforms to the Native Chieftain System." *Journal of Asian Studies* 56.1 (1997): 47–74.

———. "Mapped and Re-mapped: Chinese Representations of the Shuixi Region during the Seventeenth Century." Paper presented at the annual meeting of the Association for Asian Studies, March 26–29, 1998.

Herodotus. *The Histories.* Edited by Walter Blanco and Jennifer Tolbert Roberts, translated by Walter Blanco. New York and London: W. W. Norton and Company, 1992.

Hevia, James. *Cherishing Men From Afar: Qing Guest Ritual and the Macartney Embassy of 1793.* Durham, North Carolina, and London: Duke University Press, 1995.

Hirth, Friedrich. *Über fremde Einflüsse in der chinesischen Kunst.* Munich and Leipzig: G. Hirth, 1896.

———. "Über die chinesischen Quellen zur Kenntniss Centralasiens unter der Herrschaft der Sassaniden etwa in der Zeit 500 bis 650." *Wiener Zeitschrift für die Kunde des Morgenlandes* 10 (1896): 225–41.

Ho Ping-ti. "In Defense of Sinicization: A Rebuttal of Evelyn Rawski's 'Reenvisioning the Qing.'" *Journal of Asian Studies* 57.1 (February 1998): 123–55.

———. *Studies on the Population of China, 1368–1953.* Cambridge, Mass.: Harvard University Press, 1959.

Hodgen, Margaret T. *Early Anthropology in the Sixteenth and Seventeenth Centuries.* Philadelphia: University of Pennsylvania Press, 1964.

Hostetler, Laura. "Chinese Ethnography in the Eighteenth Century: Miao Albums of Guizhou Province." Ph.D. diss., University of Pennsylvania, 1995.

———. "Qing Connections to the Early Modern World: Ethnography and Cartography in Eighteenth-Century China." *Modern Asian Studies* 34.3 (2000): 623–62.

———. "Representation and Empire: Mapping the Qing as a Colonial Enterprise." Paper presented at the annual meeting of the Association for Asian Studies, March 26–29, 1998.

Huang Qing zhigong tu [Qing Imperial Illustrations of Tributaries]. *Qinding siku quanshu* edition, 1761.

Hucker, Charles O. *A Dictionary of Official Titles in Imperial China.* Stanford: Stanford University Press, 1985.

Hummel, Arthur. "Atlases of Kwangtung Province." *Annual Report of the Librarian of Congress for the Fiscal Year Ended June 30, 1938,* 229–31. Washington, D.C.: U.S. Government Printing Office, 1939.

———. "Astronomy and Geography in the Seventeenth Century." *Annual Report of the Librarian of Congress for the Fiscal Year Ended June 30, 1938,* 226–28. Washington, D.C.: U.S. Government Printing Office, 1939.

———. "Beginnings of World Geography in China." *Annual Report of the Librarian of Congress for the Fiscal Year Ended June 30, 1938,* 224–26. Washington, D.C.: U.S. Government Printing Office, 1939.

———. "Eastern Barbarians." *Annual Report of the Librarian of Congress for the Fiscal Year Ending June 30, 1935,* 192–93. Washington, D.C.: U.S. Government Printing Office, 1936.

———. "A Ming Encyclopedia. . . ." *Annual Report of the Librarian of Congress for the Fiscal Year Ended June 30, 1940,* 650–52. Washington, D.C.: U.S. Government Printing Office, 1941.

———. "A View of Foreign Countries in the Ming Period." *Annual Report of the Librarian of Congress for the Fiscal Year Ended June 30, 1940,* 167–69. Washington, D.C.: U.S. Government Printing Office, 1941.

Hummel, Arthur W., ed. *Eminent Chinese of the Ch'ing Period (1644–1912).* 2 vols. Taipei: Cheng Wen Publishing Company, 1970. Reprint of Library of Congress Publication, 1943, 1944.

Jacob, Margaret C. "Thinking Unfashionable Thoughts, Asking Unfashionable Questions." Review of *The Measure of Reality: Quantification and Western Society, 1250–1600,* by Alfred W. Crosby. *The American Historical Review* 105.2 (April 2000): 494–500.

Jaeger, F. "Über chinesische Miaotse-Albums." *Ostasiatische Zeitschrift* (Berlin) 4 (1916): 266–83, and 5 (1917): 81–89.

———. Review of Chiu Chang-Kong, "Die Kultur der Miao-tse . . . *"Orientalistische Literaturzeitung* 1939, no. 4., 250–52.

Jenks, Robert D. *Insurgency and Social Disorder in Guizhou: The "Miao" Rebellion, 1854–1873.* Honolulu: University of Hawaii Press, 1994.

Jiang Yingliang, ed. *Zhongguo minzu shi* [A History of China's Nationalities]. Peking Minzu chubanshe, 1990.

Kahn, Harold L. *Monarchy in the Emperor's Eyes: Image and Reality in the Ch'ien-lung Reign.* East Asian Series, no. 59. Cambridge, Mass.: Harvard University Press, 1971.

Kaltenmark, Maxime. "Chine (Périphérie de la). Mythes et legendes relatifs aux Barbares et au pays de Chou." In *Dictionnaire des mythologies et des religions, des sociétés traditionnelles et du monde antique,* 159–61. Paris: Flammarion, 1981.

"Kangxi Map Atlas" *(Huangyu quanlan tu).* 1721. British Library, c.11.d.15.

Kangxi yu Luoma shijie guanxi wenshu [Collection of Facsimile Documents Relative to Kangxi and the Roman Legations]. *Zhongguo shixue congshu,* vol. 23. Taipei: Taiwan xuesheng shuju, 1974.

Kessler, Lawrence D. *Kang-Hsi and the Consolidation of Ch'ing Rule, 1661–1684.* Chicago and London: University of Chicago Press, 1976.

Ko, Dorothy. *Teachers of the Inner Chambers: Women and Culture in Seventeenth-Century China.* Stanford: Stanford University Press, 1994.

Konvitz, Josef. *Cartography in France, 1660–1848: Science, Engineering, and Statecraft.* Chicago: University of Chicago Press, 1987.

———. "The Nation-State, Paris and Cartography in Eighteenth- and Nineteenth-Century France." *Journal of Historical Geography* 16.1 (1990): 3–16.

Kuhn, Philip. *Soulstealers: The Chinese Sorcery Scare of 1768.* Cambridge, Mass.: Harvard University Press, 1990.

Lach, Donald F., and Edwin J. Van Kley. *Asia in the Making of Europe,* vol. 3., *A Century of Advance,* bk. 4, *East Asia.* Chicago: University of Chicago Press, 1993.

Lavely, William, and R. Bin Wong. "Revising the Malthusian Narrative: The Comparative Study of Population Dynamics in Late Imperial China." *Journal of Asian Studies* 57.3 (August 1998): 714–48.

LeBar, Frank M., Gerald C. Hickey, and John K. Musgrave. *Ethnic Groups of Mainland Southeast Asia.* New Haven: Human Relations Area Files, 1964.

Lee, James Z. *The Political Economy of a Frontier: Southwest China, 1250–1850.* Harvard East Asian Monographs, 190. Cambridge, Mass.: Harvard University Press, 2000.

Lemoine, Jacques, and Chiao Chien, eds. *The Yao of South China: Recent International Studies.* Paris: Pangu, Editions de L'A.F.E.Y., 1991.

Leonard, Jane Kate. *Wei Yuan and China's Rediscovery of the Maritime World.* Cambridge, Mass.: Council on East Asian Studies, Harvard University, 1984.

Lewis, Martin W., and Karen E. Wigen. *The Myth of Continents: A Critique of Meta-Geography.* Berkeley: University of California Press, 1997.

Li Xiaocong. *A Descriptive Catalogue of Pre-1900 Chinese Maps Seen in Europe.* Peking: Guoji wenhua chubangongsi, 1996.

Li Zongfang. *Qian ji* [Record of Guizhou], 1834. *Wenyinglou yudi congshu,* compiled by Hu Sijing, 1908.

Lin Yüeh-hwa. "The Miao-man Peoples of Kweichow." *Harvard Journal of Asiatic Studies* 5 (1940): 261–345.

Liščák, Vladimír. "'Miao Albums': Their Importance and Study." *Český Lid* 78.2 (1991): 96–101.

———. "'Miaoská alba' jako etnografický pramen" ['Miao Albums' Viewed as an Ethnographical Source]. Vols. 1–2. Praha, UEF ČSAV, 1990.

Litzinger, Ralph, A. "Making Histories: Contending Conceptions of the Yao Past." In *Cultural Encounters on China's Ethnic Frontiers,* edited by Stevan Harrell. Seattle and London: University of Washington Press, 1995.

———. "Memory Work: Reconstituting the Ethnic in Post-Mao China." *Cultural Anthropology* 13.2 (1998): 224–55.

———. *Other Chinas: The Yao and the Politics of National Belonging.* Durham and London: Duke University Press, 2000.

———. "Reimagining the State in Post-Mao China." In *Cultures of Insecurity: States, Communities, and the Production of Danger,* edited by Jutta Weldes, et al. Minneapolis and London: University of Minnesota Press, 1999.

Liu, Chungshee H. "The Miao Tribes." Manuscript, 1930.

Liu Xian. "Miao tu gai lüe." *Guoli shandong daxue kexue congkan* 2 (1933): 25–37.

Lockhart, William. *The Medical Missionary in China: A Narrative of Twenty Years' Experience.* London: Hurst and Blackett, 1861.

———. "On the Miaotsze, or Aborigines of China." *Transactions of the Ethnological Society of London,* n.s., 1 (1861).

Lombard-Salmon, Claudine. *Un exemple d'acculturation chinoise: La province du Guizhou au XVIIIe siècle.* Paris: École Française d'Extrême-Orient, 1972.

Luk, Bernard Hung-Kay. "A Study of Giulio Aleni's Chih-fang Wai Chi." *Bulletin of the School of Oriental and African Studies, University of London* 40.1 (1977): 58–84.

———. "Thus the Twain Did Meet? The Two Worlds of Giulio Aleni." Ph.D. diss., Indiana University, 1977.

Ma Ruheng and Ma Dazheng, eds. *Qingdai bianjiang kaifa yanjiu* [Research into the Development of Qing Dynasty Frontiers]. Zhongguo shehui kexue chubanshe, 1990.

de Mailla, Joseph-Anne-Marie de Moyriac. *Histoire générale de la Chine; ou, Annales de cet empire: Traduites du Tong-kien-kang-mou.* 11 vols. Paris, 1777–1785. Reprint, Taipei: Ch'eng-Wen Publishing Company, 1967.

Mair, Victor H. "The Book of Good Deeds: A Scripture of the Ne People." In *Religions of China in Practice,* edited by Donald S. Lopez, Jr. Princeton: Princeton University Press, 1996.

———. "Language and Ideology in the Written Popularizations of the Sacred Edict." In *Popular Culture in Late Imperial China,* edited by David Johnson, Andrew Nathan, and Evelyn S. Rawski, 325–59. Berkeley: University of California Press, 1985.

Mann Jones, Susan. "Hung Liang-Chi (1746–1809): The Perception and Articulation of Political Problems in Late Eighteenth Century China." Ph.D. diss., Stanford University, 1972.

Mann Jones, Susan, and Philip A. Kuhn. "Dynastic Decline and the Roots of Rebellion." In *The Cambridge History of China,* vol. 10, pt. 1, 107–62.

Marcus, George E., and Michael M. J. Fischer. *Anthropology as Cultural Critique: An Experimental Moment in the Human Sciences.* Chicago: University of Chicago Press, 1986.

Mathieu, Remi. *Étude sur la mythologie et l'ethnologie de la Chine ancienne.* Paris: Collège de France, Institut des Hautes Études Chinoises, 1983.

Millward, James A. "'Coming Onto the Map': 'Western Regions' Geography and Cartographic Nomenclature in the Making of Chinese Empire in Xinjiang." *Late Imperial China* 20.2 (December 1999): 61–98.

———. "Mapping Land and History: Qing Depictions of Xinjiang/the Western Regions." Conference paper presented at the annual meeting of the Association for Asian Studies, March 12–16, 1997.

———. "Qing Xinjiang and the Trouble with Tribute." Paper presented at the 109th annual meeting of the American Historical Association, Chicago, January 5–9 1995.

Miaozu jianshi. Guiyang: Guizhou minzu chubanshe, 1985.

Mish, John. "Creating an Image of Europe for China: Alien's *Hsi-fang ta-wen.*" *Monumenta Serica* 23 (1964): 1–87.

Moule, Arthur Christopher. "An Introduction to the *I yü t'u chih* or "Pictures and Descriptions of Strange Nations" in the Wade Collection at Cambridge." *T'oung Pao* 26 (1930): 179–88.

Naquin, Susan, and Evelyn S. Rawski. *Chinese Society in the Eighteenth Century.* New Haven and London: Yale University Press, 1987.

The National Cyclopaedia of American Biography. University Microfilms, 1967; copyright 1891 by James T. White & Co.

Needham, Joseph. "Geography and Cartography." In *Science and Civilisation in China,* vol. 3, 497–590. Cambridge: Cambridge University Press, 1959.

Neumann, Carl Friedrich. "Die Urbevölkerung einer Provinzen des chinesischen Reiches." In *Asiatische Studien,* 35–120. Leipzig: J. A. Barth, 1937.

Niu Pinghan, ed. *Qingdai zhengqu yange zongbiao* [A Comprehensive Chart of the Evolution of Qing Dynasty Administrative Regions]. Peking: Zhongguo ditu chubanshe, 1990.

Paludan, Anne. "Foreigners along the Spirit Road at the Northern Song Imperial Tombs." Paper given at the Mid-Atlantic Regional Conference of the Association of Asian Studies, October 31–November 1, 1992.

Park, Nancy. "Imperial Tribute and the Rise of Official Corruption in Mid-Qing China, 1735–1796." Paper presented at the 109th annual meeting of the American Historical Society, January 5–9, 1995.

Perdue, Peter C. "Boundaries, Maps, and Movement: Chinese, Russian, and Mongolian Empires in Early Modern Central Eurasia." *International History Review* 20.2 (June 1998): 263–86.

———. "A Frontier View of Chineseness." Paper presented at the Workshop on Renegotiating the Scope of Chinese Studies, Santa Barbara, California, March 13, 2000.

———. "Military Mobilization in Seventeenth and Eighteenth-Century China, Russia, and Mongolia." *Modern Asian Studies* 30.4 (1996): 757–93.

Playfair, G. M. H. "The Miaotzu of Kweichou and Yunnan from Chinese Descriptions." *The China Review* 5.2 (1876–1877): 92–108.

Polachek, James M. *The Inner Opium War.* Cambridge, Mass.: Council on East Asian Studies, Harvard University, 1992.

Pomeranz, Kenneth. *The Great Divergence: Europe, China, and the Making of the Modern World Economy.* Princeton: Princeton University Press, 2000.

Postnikov, Alexei V. "Outline of the History of Russian Cartography." In *Regions: A Prism to View the Slavic-Eurasian World: Towards a Discipline of "Regionology,"* edited by Kimitaka Matsuzato, 1–49. Sapporo, Japan: Slavic Research Center, Hokkaido University, 2000.

Pratt, Mary Louise. *Imperial Eyes: Travel Writing and Transculturation.* London and New York: Routledge, 1992.

Rabinow, Paul. "Representations Are Social Facts: Modernity and Post-Modernity in Anthropology." In *Writing Culture,* edited by James Clifford and George E. Marcus, 234–61. Berkeley: University of California Press, 1986.

Ramsey, Robert S. *The Languages of China.* Princeton: Princeton University Press, 1987.

Rawski, Evelyn Sakakida. *Education and Popular Literacy in Ch'ing China.* Ann Arbor: University of Michigan Press, 1979.

———. *The Last Emperors: A Social History of Qing Imperial Institutions.* Berkeley, Los Angeles, and London: University of California Press, 1998.

———. "Presidential Address: Reenvisioning the Qing: The Significance of the Qing Period in Chinese History." *Journal of Asian Studies* 55.4 (November 1996): 829–50.

Ripa, Matteo. "A Map of China and the Surrounding Lands Based on the Jesuit Survey of 1708–1716." British Library. K.top.116.15, 15a, 15b.

————. *Memoirs of Father Ripa*. Translated by Fortunato Prandi. London: John Murray, 1844.

Rockhill, William W., and Friedrich Hirth, trans. *Chao Ju-Kua: His Work on the Chinese and Arab Trade in the Twelfth and Thirteenth Centuries, Entitled Chu-fan chih*. New York: Paragon Book Reprint Corp., 1966.

Rohan-Csermak, G. De. "La première apparition du terme 'ethnologie.'" *Ethnologia Europaea* 1.3 (1967): 170–84.

Rosso, Antonio Sisto, O.F.M. *Apostolic Legations to China of the Eighteenth Century*. South Pasadena: P.D. and Ione Perkins, 1948.

Rowe, William T. "Education and Empire in Southwest China: Ch'en Hung-mou in Yunnan, 1733–38." In *Education and Society in Late Imperial China, 1600–1800*, edited by Benjamin Elman and Alexander Woodside, 417–57. Berkeley, Los Angeles, and London: University of California Press, 1994.

————. *Hankow: Commerce and Society in a Chinese City, 1796–1889*. Stanford: Stanford University Press, 1984.

Ruey Yih-fu. "A Study of the Miao People." In *Symposium on Historical, Archaeological, and Linguistic Studies on Southern China, South-East Asia, and the Hong Kong Region*, edited by F. S. Drake. Hong Kong: Hong Kong University Press, 1967.

————. ed. *Miao luan (man) tu ce* [Eighty-two Aboriginal Peoples of Kweichow Province in Pictures]. Taipei: Academia Sinica, Institute of History and Philology, 1973.

————. *Fan miao hua ce* [Sixteen Aboriginal Peoples of Kweichow Province in Pictures]. Taipei: Academia Sinica, Institute of History and Philology, 1973.

Said, Edward. *Orientalism*. New York: Vintage Books, 1979.

Savina, F. M. *Histoire des Miao*. Hong Kong: Nazareth, 1930.

Schein, Louisa. "Gender and Internal Orientalism in China." *Modern China* 23.1 (January 1997): 69–98.

————. *Minority Rules: The Miao and the Feminine in China's Cultural Politics*. Durham and London: Duke University Press, 2000.

Secret Palace Memorials of the Ch'ien Lung Period. Taipei: National Palace Museum, 1982.

Shepherd, John. *Statecraft and Political Economy on the Taiwan Frontier, 1600–1800*. Stanford: Stanford University Press, 1993.

Shin, Leo Kwok-Yueh. "The Culture of Travel Writings in Late Ming China." Conference paper presented at the annual meeting of the Association for Asian Studies, March 12–16, 1997.

————. "Tribalizing the Frontier: Barbarians, Settlers, and the State in Ming South China." Ph.D. diss., Princeton University, 1999.

Shirley, Rodney W. *The Mapping of the World: Early Printed Maps, 1472–1700*. London: The Holland Press, 1993.

Sivin, Nathan, and Gari Ledyard. "Introduction to East Asian Cartography." In *The History of Cartography*, vol. 2, bk. 2, *Cartography in the Traditional East and Southeast Asian Societies*, edited by J. B. Harley and David Woodward, 23–31. Chicago and London: University of Chicago Press, 1994.

Smith, Kent C. "Ch'ing Policy and the Development of Southwest China: Aspects of Ortai's Governor-Generalship, 1726–1731." Ph.D. diss., Yale University, 1970.

Smith, Richard, J. *Chinese Maps: Images of "All Under Heaven."* Hong Kong, Oxford, and New York: Oxford University Press, 1996.

————. "Mapping China's World: Cultural Cartography in Late Imperial Times." In

Landscape, Culture and Power in Chinese Society, edited by Yeh Wen-hsin. Berkeley: University of California, Center for East Asian Studies, 1998.

Song Guangyu, ed. *Huanan bianjiang minzu tulu.* Taipei: The National Central Library and the Institute of History and Philology, Academia Sinica, 1992.

Song Zhaolin. "Qingdai Guizhou minzu de shenghuo huajuan." *Guizhou wenwu* (1987): 33–38.

———. "Qingdai Guizhou shaoshu minzu de fengsu hua." *Wenwu yuekan* (1988): 82–90.

Spence, Jonathan D. *Emperor of China, Self-Portrait of K'ang-hsi.* New York: Vintage Books, 1975.

———. "Matteo Ricci and the Ascent to Peking." In *East Meets West: The Jesuits in China, 1582–1773,* edited by Charles E. Ronan, S.J., and Bonnie B. C. Oh, 3–18. Chicago: Loyola University Press, 1988.

———. *The Memory Palace of Matteo Ricci.* New York: Penguin Books, 1984.

———. *To Change China: Western Advisers in China, 1620–1960.* Boston and Toronto: Little, Brown and Company, 1969.

Spenser, J. E. "Kueichou: An Internal Chinese Colony." *Pacific Affairs* 13 (1940): 162–72.

Sturtevant, William C. "The Sources for European Imagery of Native Americans." In *New World of Wonders: European Images of the Americas, 1492–1700,* edited by Rachel Doggett, with Monique Hulvey and Julie Ainsworth, 25–33. Washington, D.C.: The Folger Shakespeare Library, 1992.

Subrahmanyam, Sanjay. "Connected Histories: Notes towards a Reconfiguration of Early Modern Eurasia." *Modern Asian Studies* 31.3 (1997): 735–62.

Szczesniak, Boleslaw. "The Atlas and Geographic Description of China: A Manuscript of Michael Boym (1612–1659)." *Journal of the American Oriental Society* 73 (1953): 65–77.

———. "Matteo Ricci's Maps of China." *Imago Mundi* 11 (1955): 127–36.

Tan Qixiang, ed. *The Historical Atlas of China.* Vol. 8, *The Qing Dynasty Period (Zhongguo lishi ditu ji Qing shiqi).* Shanghai: Cartographic Publishing House, 1987.

Tanaka, Stephen. *Japan's Orient: Rendering Pasts into History.* Berkeley, Los Angeles, and London: University of California Press, 1993.

Teng, Emma. "Mapping Emptiness: Visual and Literary Representations of Taiwan's Wilderness." Paper presented at the annual meeting of the Association for Asian Studies, March 26–29, 1998.

Teng, Ssu-yü. *Ssu-ma Ch'ien yü Hsi-lo-to-te chih pi chiao (Szu-ma Ch'ien and Herodotus). Bulletin of the Institute of History and Philology Academia Sinica,* vol. 27, *Studies Presented to Hu Shih on His Sixty-fifth Birthday.* Taiwan: Academia Sinica, 1957.

———. *Herodotus and Ssu-ma Ch'ien: Two Fathers of History.* Rome: Ismeo, 1961.

Thongchai Winichakul. *Siam Mapped: A History of the Geo-Body of a Nation.* Honolulu: University of Hawaii Press, 1994.

Tian Rucheng. *Yanjiao jiwen,* 1560. *Yingyin wenyuange siku quanshu* edition. Taipei: Taiwan shangwu yinshuguan reedition, 1983.

Tian Wen. *Qian shu,* 1690. *Qiannan congshu jicheng* edition. Guiyang: Wentong shuju, [1936?].

Toby, Ronald P. "The Indianness of Iberia and Changing Japanese Iconographies of Other." In *Implicit Understandings: Observing, Reporting, and Reflecting on the Encounters between Europeans and Other Peoples in the Early Modern Era,* edited by Stuart B. Schwartz, 323–51. Cambridge: Cambridge University Press, 1994.

————. "The Race to Classify." *Anthropology Newsletter* 39.4 (April 1998): 55–56.

Tooley, R. V. *Maps and Map-Makers.* 4th ed. New York: Bonanza Books, 1970.

Vanderstappen, Harrie, S.V.D. "Chinese Art and the Jesuits in Peking." In *The Jesuits in China, 1582–1773,* edited by Charles E. Ronan, S.J., and Bonnie B. C. Oh, 103–26. Chicago: Loyola University Press, 1988.

Vaughan, Alden T. "People of Wonder: England Encounters the New World's Natives." In *New World of Wonders: European Images of the Americas, 1492–1700,* edited by Rachel Doggett, with Monique Hulvey and Julie Ainsworth, 11–23. Washington, D.C.: The Folger Shakespeare Library, 1992.

Verbiest, Ferdinand. *Kun yu quan tu.* 1674. The British Library, Maps 183.p.4[1].

Waley-Cohen, Joanna. "China and Western Technology in the Late Eighteenth Century." *American Historical Review* 98.5 (1993): 1525–44.

————. "Commemorating War in Eighteenth-Century China." *Modern Asian Studies* 30.4 (1996): 869–99.

————. "Religion, War, and Empire-Building in Eighteenth-Century China." *International History Review* 20.2 (June 1998): 336–52.

————. *The Sextants of Beijing: Global Currents in Chinese History.* New York and London: W. W. Norton and Company, 1999.

Widmer, Ellen. "The Huanduzhai of Hangzhou and Suzhou: A Study in Seventeenth-Century Publishing." *Harvard Journal of Asiatic Studies* 56 (June 1996): 77–122.

Wiens, Herold J. *China's March toward the Tropics.* Hamden, Conn.: Shoe String Press, 1954.

Williams, Samuel Wells. "Notices of the Miao Tsz, or, Aboriginal Tribes Inhabiting Various Highlands in the Southern and Western Provinces of China Proper." *Chinese Repository* 14.3 (1845): 105–15.

Wills, John E. Jr. "Maritime Asia, 1500–1800: The Interactive Emergence of European Domination." *American Historical Review* 98.1 (1993): 83–105.

————. "Tribute, Defensiveness, and Dependency: Uses and Limits of Some Basic Ideas about Mid-Qing Dynasty Foreign Relations." *American Neptune* 48 (1988): 225–29.

Wolf, C. *Histoire de l'Observatoire de Paris de sa fondation à 1793.* Paris: Gauthier-Villars, 1902.

Wolter, John A., and Ronald E. Grim, eds. *Images of the World: The Atlas through History.* Washington, D.C.: Library of Congress, Center for the Book, 1997.

Wong, Roy Bin. *China Transformed: Historical Change and the Limits of European Experience.* Ithaca: Cornell University Press, 1997.

Wood, Frances. *Did Marco Polo Go to China?* Boulder, Colo.: Westview Press, 1995.

Woodward, David. "Reality, Symbolism, Time, and Space in Medieval World Maps." *Annals of the Association of American Geographers* 75.4 (1985): 510–21.

Xuan he hua pu. In *Xue jin tao yuan,* vol. 22.

Yee, Cordell D. K. "A Cartography of Introspection: Chinese Maps as Other Than European." *Asian Art* (fall 1992): 28–47.

————. "Reinterpreting Traditional Chinese Geographical Maps." In *The History of Cartography,* vol. 2, bk. 2, *Cartography in the Traditional East and Southeast Asian Societies,* edited by J. B. Harley and David Woodward, 35–70. Chicago and London: University of Chicago Press, 1994.

————. "Traditional Chinese Cartography and the Myth of Westernization." In *The*

History of Cartography, vol. 2, bk. 2, *Cartography in the Traditional East and Southeast Asian Societies,* edited by J. B. Harley and David Woodward, 170–202. Chicago and London: University of Chicago Press, 1994.

Yuan Ke, ed. *Zhongguo shenhua chuanshuo cidian.* Taipei: Huashe chubanshe, 1987.

Zögner, Lothar. *China cartographica: Chinesische Kartenschätze und europäische Forschungsdokumente.* Berlin: Staatsbibliothek Preussischer Kulturbesitz, 1983.

GLOSSARY

Angshui	邛水
Anshun	安順
Badan	八丹
Bafan Miao	八番苗
Bafanzi	八番子
Bai (surname)	白
Bai chuan shu zhi	白川書志
Bai Erzi	白兒子
Bai Ezi	白頟子
Bai Longjia	白龍家
Bai Luoluo	白猓玀
Bai Miao	白苗
bai Miao tu	白苗圖
"Bai Miao tu"	白苗圖
"Bai Miao tuyong"	白苗圖永
Bai Zhongjia	白狆家
Bangshui	邦水
Bazhai	八寨
Bazhai Hei Miao	八寨黑苗
bian	邊
Bianqiao	偏橋
Bin Yisheng	蟫衣生
Boren	僰人
Bozhou	播州
Bulong Zhongjia	補籠狆家
buluo	部落
Buyi	布依

Caijia	蔡家
Caodi	曹滴
Caodidong	曹滴洞
Celeng	策楞
Cengzhu Longjia	曾竹龍家
changhao	嗜好
Changshun	長順
Chen Hao	陳浩
Chen Shen	陳詵
Chen Zongang	陳宗昂
Chezhai Miao	車寨苗
Chuang Chi-fa (Zhuang Jifa)	莊吉發
chunpu	淳樸
Ci Yuan	辭源
Da Ming yi tong zhi	大明一統志
Dading	大定
Danjiang	丹江
Danping	丹平
Danren	蜑人
Danxing	丹行
Datou Longjia	大頭龍家
Daya Gelao	打牙犵狫
"Dian yi tu shuo"	滇夷圖説
"Diansheng yi xi yi nan yiren tushuo"	滇省迤西迤南夷人圖説
dili	地理
Dingfan	定番
Dingying	頂營
Dizhai	底寨
Dong	峒
dong lu	東路
Dong (cave) Miao	硐(洞)苗
Dong (east) Miao	東苗
Dong Miao Xi Miao	東苗西苗
Dong yi kao lüe	東夷考畧
Dongjia	峒家
Dongjia Miao	峒家苗
Dongren	峒人
Dongting	洞庭
Dongzai	洞崽
Dongzai Miao	洞崽苗

Duanqun Miao	短裙苗
Du shu min qiu ji	讀書敏求集
Dushan	獨山
Duyun	都勻
Eguo bing ying	俄國兵營
fan min	番民
fan ren	番人
fan zu	蕃族
Fangguo shitu	方國使圖
fangzhi	方志
"Fanshe caifeng tukao"	番社採風圖考
Fuheng	傅恆
Fujian	福建
gaitu guiliu	改土歸流
Gao Ru	高儒
Gaopo Miao	高坡苗
Ge Zhuang	犵獞
Gedou	犵兜
Gelao	犵狫, 犵老
Gellao	犵獠
Gouer Longjia	狗耳龍家
gu	蠱
guan	館
guang	獷
Guang yu tu	廣輿圖
Guangdong	廣東
Guangxi	廣西
Guangxu	光緒
guanshu	官署
Guangshun	廣順
Guiding	貴定
guigan	鬼竿
Guiyang	貴陽
Guizhou	貴州
"Guizhou Miaotu"	貴州苗圖
Guizhou tongzhi	貴州通志
"Guizhou zhu yi tu"	貴州諸夷圖
Guizhu	鬼主
Gujin tushu jicheng	古今圖書集成
Gulin Miao	谷藺苗

guo	國
Guo Qingluo	郭青螺
Guo Zizhang	郭子章
Guoquan Gelao	鍋圈犵狫
Guyang	牯羊
Guzhou	古州
han (fierce)	悍
Han (dynasty, ethnicity)	漢
hanhua	漢化
hanyu	漢語
hao	號
He Changgeng	賀長庚
Hei Gelao	黑犵狫
Hei Jiao Miao	黑脚苗
Hei Lou Miao	黑樓苗
Hei Luoluo	黑猓玀
Hei Miao	黑苗
Hei Shan Miao	黑山苗
Hei Sheng Miao	黑生苗
Hei Zhongjia	黑狆家
Hong Gelao	紅犵狫
Hong Miao	紅苗
Hongzhou	洪州
Hongzhou Miao	洪州苗
Hongwu	洪武
Hua Gelao	花犵狫
Hua Miao	花苗
Huai (river)	淮(河)
Huang Qing zhigong tu	皇清職貢圖
Huangyu quanlan tu	皇輿全覽圖
Hu'er	湖耳
Huguang	湖廣
Hui min	回民
Hulu Miao	葫蘆苗
Hunan	湖南
ji	記
Ji (surname)	姬
jia	家
jian lu	間路
Jianding Miao	尖頂苗

Jianfa Gelao	剪髮犵狫
Jiang Dejing	蔣德璟
Jiangnan	江南
Jiangxi	江西
Jiantou Gelao	剪頭犵狫
jiaofu	勦撫
Jiaqing	嘉慶
jiejian	接見
Jin (dynasty)	晉
Jin (surname)	金
Jinbing	錦屏
jingli	經歷
Jingzhou	靖州
jinshi	進士
Jinzhu	金筑
Jiugu Miao	九股苗
Jiuming Jiuxing Miao	九名九姓苗
ju	狙
juan	卷
juren	舉人
Kaili	凱里
Kangxi	康熙
kaozheng	考證
Kayou	卡猶
Kayou Zhongjia	卡猶狆家
Kemeng	克孟
Kemeng Guyang Miao	克孟牿羊苗
Kunyu tushuo	坤輿圖説
Lang Shining	郎世寧
Langci Miao	郎慈苗
li (measure of distance)	里
li (rustic)	俚
Li (surname)	黎
Li (ethnicity)	黎
Li Miao	里苗
Li Minzi	里民子
Li Zhizao	李之藻
Li Zongfang	李宗昉
Liang (dynasty)	梁
Liangzhai	亮寨

Liaoren	獠人
lie zhuan	列傳
Lin Yüe-hwa (Lin Yaohua)	林耀華
Lingjia	狑家
Lingjia Miao	狑家苗
Liping	黎平
Liren	黎人
"Liren fengsu tu"	黎人風俗圖
Liu Ezi	六額子
Liu Jin	劉錦
Liudong Yiren	六峒夷人
Long (ethnicity)	龍
Long (surname)	龍
Longli	龍里
Louju Miao	樓居苗
Lulu	鹿盧
Luo Jihuo	羅濟火
Luodian	羅甸
Luogui	羅鬼
Luohan Miao	羅漢苗
Luohei	猓黑
Luoluo	猓玀
lusheng	蘆笙
Ma (surname)	麻
Ma Duanlin	馬端臨
Ma Huan	馬歡
Madeng Longjia	馬蹬龍家
Maiye Miao	賣爺苗
man	蠻
Man Liao	蠻獠
"Man Liao tushuo"	蠻獠圖說
Manren	蠻人
Mao Ruizheng	茅瑞徵
Maren	馬人
Miandian	緬甸
miao (sprout)	苗
Miao (ethnicity)	苗
Miao liao	苗獠
Miao man	苗蠻
Miao man tu	苗蠻圖

Miao luan (man) tu ce	苗繺(蠻)圖冊
"Miao man tu ce ye"	苗蠻圖冊頁
Miao sheng	苗生
Miaomin	苗民
min	民
Ming	明
mu	目
Mulao	木狫
Nan Zhao	南詔
nei wai miao yi	內外苗夷
neidi	內地
Ninggu	寧谷
Nong (ethnicity)	儂
Nong Miao	儂苗
Nüguan	女官
Ortai	鄂爾泰
Ou	歐
Pan Hu	盤瓠
Pei Xiu	裴秀
Peking (Beijing)	北京
piao duo wei chang	票奪為常
Pingfa Miao	平伐苗
Pingfa si	平伐司
Pingfa si Miao	平伐司苗
Pingyuan	平遠
Pingyue	平越
Pipao Gelao	披袍犵狫
pu	樸
Pu'an	普安
Puliao	僕獠
Qian ji	黔記
"Qian Miao tushuo"	黔苗圖說
Qian shu	黔書
Qian Zeng	錢曾
qiang han hao dou	強悍好鬥
Qianlong	乾隆
"Qiannan Miao man tushuo"	黔南苗蠻圖說
Qiannan zhifang jilüe	黔南職方紀略
Qianxi	黔西
qimin	齊民

qing (blue)	青
Qing (dynasty)	清
Qing Miao	青苗
Qing Zhongjia	青狆家
Qingjiang	清江
Qingjiang Miao	清江苗
Qingjiang Zhongjia	清江狆家
Ranjia	冉家
Ranjia Man	冉家蠻
ren	人
renwu	人物
San cai tu hui	三才圖會
San Miao	三苗
Shan hai jing	山海經
Shanghai	上海
Shanxi	陝西
She Xiang	奢香
sheng	生
sheng	笙
Sheng Miao	生苗
Sheng Miao Hong Miao	生苗紅苗
Sheng Miao jie	生苗界
shengyu	聖諭
Shengyu guangxun	聖諭廣訓
shi (power)	勢
Shi (surname)	石
Shiji	史記
Shibing	施秉
Shiqian	石阡
shu (i.e. *Qian shu*)	書
shu (tame, cooked)	熟
Shui Gelao	水犵狫
Shui, Yang, Ling, Dong, Yao, Zhuang (liu zhong)	狏, 犲, 狑, 狪, 猺, 獞 (六種)
Shuijia	水家
Shuijia Miao	水家苗
shun liang	馴良
si	司
Si Longjia	四龍家
Sichuan	四川

Siku quanshu	四庫全書
Sima Qian	司馬遷
Sinan	思南
Sizhou	思州
Song (dynasty)	宋
Song Zhaolin	宋兆麟
Songjia	宋家
Su Jiyu	蘇及寓
Sui (dynasty)	隋
"Taifan tushuo"	臺番圖説
Taihe	泰和
Taiwan	臺灣
Tang (dynasty)	唐
Tanqi	潭溪
Tian Rucheng	田汝成
Tianzhu	天柱
"Taiwan neishan fandi fengsu tu"	臺灣內山番地風俗圖
Taiwan wenxian congkan	臺灣文獻叢刊
Tan Qixiang	譚其驤
ting guanfu yueshu	聽官府約束
tiyao	提要
Tong dian	通典
Tongren	銅仁
tongzhi	通志
tu	圖
Tu Gelao	土犵狫
Tujia	土家
Tujue suo chu fengsu shi	突厥所出風俗事
tuntian	屯田
Turen	上人
tushi	圖式
tusi	土司
tuxiang	圖像
tuyang	圖樣
wai yi fan zhong	外夷番眾
Waiguo tu	外國圖
wan hao zhi ju	玩好之具
Wang (surname)	王
Wang Lüjie	王履階
Wang Pan	王泮

Wang Qi	王圻
Wanghui tu	王會圖
Wanyong zhengzong buqiuren quanbian	萬用正宗不求人全編
Weiqing	威清
Wen (surname)	文
Wu	武
Wu (hou)	武(侯)
Wu (surname)	吳
Wuchuan	婺川
xi fan	西番
xi lu	西路
Xi Miao	西苗
Xi yu tu ji	西域圖記
Xianbin lu	咸賓錄
xianghua	向化
xiao si kou	小司寇
Xibao	西堡
Xie Sui	謝遂
Xifang dawen	西方答問
Xifang yaoji	西方要紀
xing	性
Xinglong	興隆
Xinhua	新化
Xintian	新添
Xiqi Miao	西溪苗
Xishan Yangdong	西山陽洞
Xu Guangqi	徐光啓
Xuan he hua pu	宣和畫譜
xuanweisi	宣慰司
xunfu	馴服
Yan Liben	閻立本
Yang (surname)	楊
Yang Delin	楊德霖
Yangbao Miao	楊保苗
Yangdong Luohan Miao	陽峒羅漢苗
Yangguang Miao	狋獷苗
Yanghuang Miao	狋獚苗
Yangzi (Changjiang)	長江
Yanhe	沿河
Yanhe youqi	沿河祐溪

Yanjiao Jiwen	炎徼紀聞
Yao	猺
Yao Miao	夭苗
Yaojia	猺家
Yaoren	猺人
Yaque Miao	鴉雀苗
Ye Xianggao	葉向高
Yelang	夜郎
Yelang hou	夜郎侯
Yetou Miao	爺頭苗
Yi (ethnicity)	彞
yi	夷
yi di	夷地
Yi yu tu zhi	異域圖志
yiguan wenwu	衣冠文物
ying	營
Ying yai sheng lan	瀛涯勝覽
yixue	義學
Yongcong	永從
Yongzheng	雍正
you	莠
yu	愚
Yu ji tu	禹跡圖
Yu Yushan	俞愉山
Yuan (dynasty)	元
Yuan ming yuan	圓明園
Yuan Shu	袁術
yue chang	月場
Yunnan	云南
"Yunnan san yi bai man tu"	云南三迤百蠻圖
Zangge	牂柯
zhai	寨
Zhang	張
Zhang Guangsi	張廣泗
Zhao Rugua	趙汝适
Zhenning	鎮寧
Zhenyuan	鎮遠
zhidu	制度
zhifang si	職方司
Zhifang waiji	職方外紀

zhigong tu	職貢圖
zhong	仲, 狆
Zhongguo	中國
Zhongjia	狆家
Zhonglin	中林
zhongtu	中土
zhou	州
zhu	竹
Zhu fan fengsu ji	諸蕃風俗記
Zhu fan zhi	諸蕃志
Zhu Wang	竹王
Zhu yi za zhi	諸夷雜誌
zhuan	撰
zhuang	憃
Zhuangren	獞人
Zhuge Liang	諸葛亮
Zhuliao	主獠
"Zhuluo xianzhi fansu tu"	諸羅縣志番俗圖
zhuo	拙
Zhushi Gelao	豬屎犵狫
Zihong	紫紅
Zijiang Miao	紫薑(姜)苗
Zunyi	遵義

INDEX